Springer-Verlag Berlin Heidelberg GmbH

Tim Clark · Rainer Koch

The Chemist's Electronic Book of Orbitals

With 90 Figures

 Springer

Dr. Tim Clark
Universität Erlangen-Nürnberg
Computer-Chemie-Centrum
Nägelsbachstraße 25
D-91052 Erlangen

Dr. Rainer Koch
Universität Oldenburg
FB Chemie
P.O. Box 25 03
D-26111 Oldenburg

Library of Congress Cataloging-in-Publication Data

Clark, Tim
  The chemist's electronic book of orbitals / Tim R. Clark,
Rainer Koch.

  1. Molecular orbitals. I. Koch, Rainer- II. Title.
  QD461 .C415 1998
  541.2'2–ddc21   98-46052

ISBN 978-3-540-63726-4   ISBN 978-3-662-13150-3 (eBook)
DOI 10.1007/978-3-662-13150-3

© Springer-Verlag Berlin Heidelberg 1999

Originally published by Springer-Verlag Berlin Heidelberg New York in 1999.

The use of general descriptive names, registered names, trademarks, etc. in this publication does not
imply, even in the absence of a specific statement, that such names are exempt from the relevant pro-
tective laws and regulations and therefore free for general use.

Typesetting: Data conversion by MEDIO, Berlin
Cover design: de'blik, Berlin

SPIN: 10633091   52/3020 5 4 3 2 1 0   Printed on acid-free paper.

Preface

For many years, Bill Jorgensen and Lional Salem's *Organic Chemist's Book of Orbitals* was the standard work on qualitative MO-theory in Erlangen. It provided the basic principles as well as MO-plots of exactly the orbitals that we needed to understand the effects that we were investigating. However, 25 years after its publication, the *Organic Chemist's Book of Orbitals* is essentially unknown to the current generation of chemistry students, although qualitative MO-theory has become a standard tool. This and the new technical possibilities made possible by the development of web browsers, VRML-viewers etc. suggested that the time is right to attempt a more modern version, this time called *The Chemist's Electronic Book of Orbitals*. The resulting "book" is a tribute to its predecessor and we hope that it will play a similar role as vanguard for a new generation of chemistry publications in this area. The written text can be read alone, but is intended to be used in conjunction with the demonstrations and VRML-objects on the CD. The orbital plots can be rotated, displayed and printed as desired. We have designed *The Chemist's Electronic Book of Orbitals* to be accessed from a standard web-browser with VRML viewer and JAVA in order to avoid the need for extra software on the CD and to make it accessible from as many hardware platforms as possible.

Many people have contributed to the success of this project. We especially thank Peter Enders at Springer for unflagging support. The entire Clark-group has devoted many coffee breaks to discussing possibilities and improvements and Jens Schamberger and Peter Gedeck developed the graphics program used to plot the MOs. Peter Comba provided us with a list of the most important ligands for transition metal complexes, but any omissions are ours, not his. It is a pleasure to thank the Deutscher Forschungsnetz-Verein (DFN) and especially Ralf Paffrath. This book was conceived as part of a DFN-financed teleteaching project and has benefitted from many stimulating discussions within this project. Above all, however, we thank Bill Jorgensen and Lionel Salem for the "one and only original" and hope that they will find our "son of Jorgensen and Salem" worthy of its predecessor.

Tim Clark and Rainer Koch
Erlangen, Oldenburg, January 1999

Contents

1 Introduction

In 1973, Bill Jorgensen and Lionel Salem published a small paperback that was to become the constant companion of many physical organic chemists for years to come. The *Organic Chemist's Book of Orbitals* was delightful in its simplicity and clarity of presentation of the concepts of group orbitals and qualitative molecular orbital theory in general. For many of us, it was symbolic of a chemical era. Sadly, it went out of print far too early. Now, 25 years later, the concepts presented by Jorgensen and Salem are as valid as ever and have advanced to become the major interpretational tool of organic and inorganic chemists. However, the technical possibilities for presenting this material have changed beyond recognition. We have therefore revived the idea of the *Organic Chemist's Book of Orbitals* and given it a new form that allows the reader (or is the "reader" now the "user"?) to rotate and zoom the molecular orbital (MO) plots, which are stored as three-dimensional objects. As in the original, the MO-plots are preceded by a short text section on the principles of qualitative MO-theory, although we have also added a short introduction to elementary symmetry considerations in order for the reader to be able to understand the symmetry designations of the individual orbitals. We have also added MO-plots for some of the more important organic compounds that are commonly used as ligands for transition metal complexes.

Chapter 2 presents the principles of the *Linear Combination of Atomic Orbitals* (*LCAO*) approximation. This is the dominant approximation of qualitative (and 99 % of quantitative) MO-theory. It has become so established that we very often lose sight of the fact that it is an approximation and there have been many controversial discussions about methods for analysing the properties of individual atoms within molecules over the last 25 years. Chapter 2 also describes the effects of the electronegativities of the individual elements and introduces the Walsh diagram approach to analysing the structures of small molecules and fragments.

Chapter 3 extends these concepts to larger molecules. The MOs of one-heavy-atom fragments can be used as group orbitals to build up the MOs of larger molecules. This concept of group orbitals often provides the basis for the interpretation of MO-effects in larger molecules. Chapter 3 then moves on to treat π-systems, which are involved in much of the reactivity that we seek to explain using MO-concepts, and hyperconjugation.

Chapter 4 extends the qualitative MO-concepts to reactions and gives a very short treatment of the Woodward-Hoffman rules. The treatment of concerted electrocyclic reations, cycloadditions and sigmatropic rearrangements provided the impetus for the adoption of MO-treatments, rather than the more traditional resonance theory, in organic chemistry. We now have a rather unsatisfactory situation in which elementary mechanistic organic chemistry is treated using valence-bond and resonance arguments (because they work and are easy to explain) but that for the above processes we change to MO-theory. There have been many heroic attempts to resolve this dichotomy in organic textbooks in the last 20 years, but an ideal solution remains elusive.

Chapter 5 gives a short introduction to molecular symmetry, symmetry elements, point groups, Mulliken symbols and character tables. It is not intended to provide a complete treatment of the subject, but rather to help the non-specialist understand the symbols and perhaps to prepare students for more specialised textbooks on symmetry. The final section of this chapter treats degenerate MOs. The MO-plots for these orbitals used moelcular structures that were slightly distorted from the full molecular symmetry so that the degenerate MOs appear in their usual forms with nodal planes corresponding to molecular mirror planes. It is, however, important to realise that real MO-calculations using the full symmetry may give different, equivalent sets of MOs.

Chapter 6 contains the MO-plots themselves. The printed version simply contains the table of the molecules available with their point groups and molecular formulae. In the CD-ROM version, this table is linked to the MO-plots themselves, which are available as *Virtual Reality Markup Language* (*VRML*) objects that can be rotated, zoomed and printed in the desired orientation. It may be necessary to install the appropriate VRML-viewer from the CD if your Web browser is not already equipped for VRML. The MO-plots themselves were all produced using AM1 semiempirical MO-theory at the AM1-optimised geometries. Generally, the differences between orbitals plotted at this level of theory and those given by high level large basis set ab initio calculations are hardly visible, so that we have chosen the fast, efficient semiempirical method that has become so popular for large organic molecules. The geometries may, however sometimes deviate significantly from those given by higher levels of theory.

We intend this edition to be the beginning of a continuous development that will take advantage of new technologies to provide more interactive multimedia features in the text and additional ways for the user to interact with the MO-plots. Time will tell whether the format that we have chosen will survive long into the future, but we have tried to make the CD-ROM as platform-independent as possible so that it can be used by PC, Mac and Unix users equally well without the need to buy additional software.

2 Linear Combination of Atomic Orbitals

The *Linear Combination of Atomic Orbitals* (*LCAO*) approximation is fundamental to many of our current models of chemistry. Both the vast majority of the calculational programs that we use, be they *ab initio*, density functional, semiempirical molecular orbital, or even some sophisticated forcefields, and our qualitative understanding of chemistry are based on the concept that the orbitals of a given molecule can be built from the orbitals of the constituent atoms. We feel comfortable with the π-*HOMO* (*H*ighest *O*ccupied *M*olecular *O*rbital) of ethylene depicted as a combination of two carbon *p*-orbitals, as shown in Fig. 2.1, although this is not a very accurate description of the electron density of this *Molecular Orbital* (*MO*). The use of the π-*A*tomic *O*rbitals (*AOs*), however, makes it easier to understand both the characteristics of the MO itself and the transformations that it can undergo during reactions.

In principle, we could build up MOs from many sorts of function that can describe an electron density probablility distribution, but we have learnt to understand combinations of AOs and to use them in our models of chemical bonding and reactivity. Indeed, if confronted with an MO that was calculated, for instance, using plane waves, most chemists would immediately translate it into a combination of AOs. In the following, we will describe the LCAO approximation and demonstrate some of the effects that are important when AOs interact with each other to form MOs.

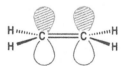

Fig. 2.1 π-HOMO of ethylene

2.1
H₂ and He₂ – The Simplest Examples

The simplest example of the LCAO approximation is the combination of two *s*-AOs to form s- (bonding) and σ*- (antibonding) MOs. This is shown in Fig. 2.2 for the dihydrogen molecule. Figure 2.2a shows the simple *orbital interaction diagram*, whereas Fig. 2.2b shows *3D-electron density contour plots* for the AOs and MOs.

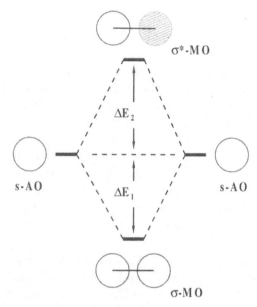

Fig. 2.2a Schematic orbital interaction diagram for a system consisting of two *s*-orbitals

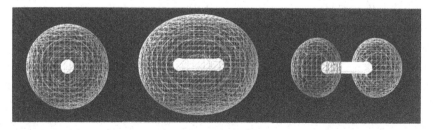

Fig. 2.2b Orbital electron density plots for the AOs and MOs shown in Fig. 2.2a. From left to right: *s*-AO, σ-MO, σ*-MO

It is useful to define the conventions used in Fig. 2.2a and all other orbital interaction diagrams in this book. The thick horizontal lines represent orbital energy levels and, although it is not explicitly shown, the diagram implies a vertical energy scale. The dashed connecting lines signify orbital interactions. Thus, the two s-orbitals, which have the same energy, interact with each other to form the σ- and σ*-MOs. The σ-MO is stabilised by ΔE_1 relative to the energy level of the two AOs and the σ*-MO is destabilised by ΔE_2. It is important to note that ΔE_1 is always slightly smaller than ΔE_2.

If we now add the electrons to the orbital interaction diagram, we find that H$_2$, with two electrons (see Fig. 2.3a), can form a closed-shell singlet state in which the electrons are paired in the stabilized σ-orbital to give a strong bond (the total energy of the two electrons is lower than in the two separated atoms). If, however, we consider He$_2$, with four electrons (see Fig. 2.3b), we must doubly occupy both the σ- and the σ*-orbitals. Because ΔE_1 is smaller than ΔE_2, this situation gives a higher energy than for the four electrons in the two separated atoms. He$_2$ thus does not form a bond.

The stabilisation and destabilisation energies, ΔE_1 and ΔE_2, depend on the *overlap* between the AOs. The closer the two atoms approach each other, the larger the overlap and the larger the energy splitting. This effect can be tested using *demonstration 1*.

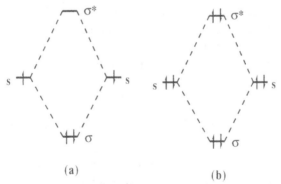

(a)   (b)

Fig. 2.3 Electron occupations for (a) H$_2$ and (b) He$_2$

Demo 1 Interactive MO scheme for H$_2$ and He$_2$; orbital contours, and energies and bonding energy for two or four electrons as a function of distance.

2.2
The Effect of Electronegativity

Electronegativity measures the tendency of an atom to accept electrons. Electronegative elements tend to accept electrons from less electronegative ($\equiv$ more *electropositive*) ones. Thus, electronegative atoms are usually negatively charged (they have more electrons than they need to equalise the core charge) and, oddly enough, electropositive ones are positively charged. In orbital terms, the AO energy levels of electronegative elements lie lower in energy than those of electropositive ones. Figure 2.4 gives the Pauling electronegativities and the electronic configuration for neutral atoms.

The effect of differing electronegativities on the MOs of the simple two *s*-orbital system shown above is shown in Fig. 2.5 for LiH.

The σ-MO, which is closer to the energy level of the *s*-AO of the more electronegative hydrogen, is composed of a larger proportion of this AO than of the higher energy lithium *s*-AO. In contrast, the situation is reversed for the σ^*-MO, which is closer in energy to the *s*-AO of the more electropositive lithium. This situation can be expressed in the LCAO expression for the two MOs:

$$\Psi_\sigma = c_1\psi_{s(H)} + c_2\psi_{s(Li)} \tag{1}$$

$$\Psi_{\sigma*} = c_3\psi_{s(H)} + c_4\psi_{s(Li)} \tag{2}$$

where Ψ_s and Ψ_{s*} are the two molecular orbitals, $y_{s(X)}$ the *s*-AO on element X and the coefficients $c_1 - c_4$ determine the relative weights of the AOs in the MOs. These coefficients are *normalised*, so that the total probabilty of finding an electron in the orbital is one. In this case this *normalisation* is defined by:

$$c_1^2 + c_2^2 = 1 \tag{3}$$

and

$$c_3^2 + c_4^2 = 1 \tag{4}$$

for the σ- and σ^*-MOs, respectively. The coefficients can be used to divide the electrons between the two atoms involved in the orbital. Because the σ-MO is doubly occupied, the number of electrons, N_H on the hydrogen atom can be defined as

$$N_H = 2 \cdot \frac{c_1^2}{c_1^2 + c_2^2} = 2 \cdot c_1^2 \tag{5}$$

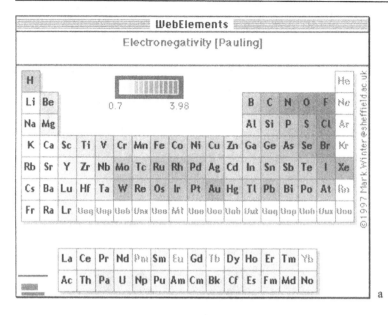

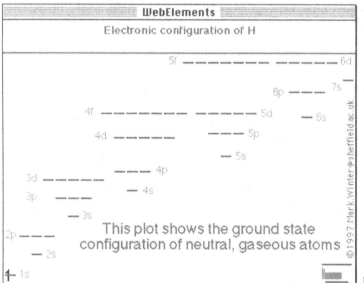

Fig. 2.4 Pauling electronegativities **a** and schematic electronic configuration **b**

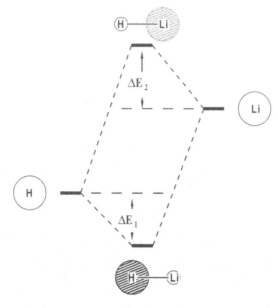

Fig. 2.5 Orbital interaction diagram for LiH, illustrating the effect of differing electronegativities. The AOs of H and Li have been drawn equally large to emphasise the contributions to the MOs

This procedure is known as a *population analysis*, and is often used to assign *net atomic charges*, which, however, have no physical meaning, but are simply defined to help interpret the electronic nature of the compound to be studied. It follows that the more electronegative atom, which has the larger coefficient in the σ-MO, also has the more negative net atomic charge. *Demonstration 2* illustrates the effect of changing the relative electronegativities of the two elements shown in Fig. 2.5 on the shapes of the MOs, the bonding energy, the *dipole moment* and the net atomic charges.

Demo 2 Interactive orbital interaction diagram between two s-AOs; MOs, dipole moment, bond energy and atomic charges as a function of bond distance and electronegativity difference.

2.3
p-Orbitals and π-Overlap

The interaction diagrams in Sects. 2.1 and 2.2 dealt with *s*-AOs. Because they are spherically symmetric, *s*-orbitals have no directional preferences. In contrast, *p*-AOs interact very differently depending on their relative directions. Two *p*-AOs can, for instance, form σ- and σ*-MOs, as shown in Fig. 2.6.

Note that if they approach each other too closely, two *p*-orbitals in a σ-MO can actually begin to overlap less well than at larger distances, as shown in Fig. 2.7.

Another important aspect of *p*-orbital overlap is its angle-dependence. Fig. 2.8 shows the progression from a pure σ-overlap to a *"banana" bond* by rotation of the two orbitals away from each other. Such bonds are found in strained compounds such as cyclopropanes (see Sect. 3.2). Further rotation of the two orbitals away from each other results in pure π-overlap, in which the two *p*-AOs are parallel to each other.

Note that the overlap between the two orbitals (assuming that the distance between their centres remains constant) is largest in the σ-situation

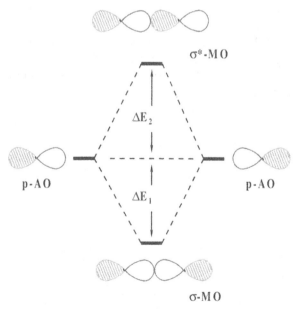

Fig. 2.6 The interaction of two *p*-AOs to give σ- and σ*-MOs

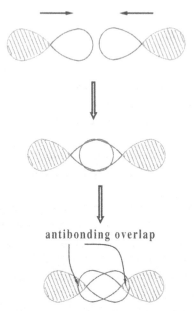

Fig. 2.7 As two *p*-orbitals approach each other in a σ-fashion (top) they reach a position of optimum overlap (centre) and, on approaching closer, one in which the total overlap is decreased by antibonding interactions outside the nodal planes of the two AOs (bottom)

and becomes smaller as the orbitals are twisted away from each other. The pure π-MO represents a minimum in the overlap. Thus, "banana" and π-orbitals lie higher in energy than their σ-counterparts. This energy difference is manifested as *strain* in organic compounds. *Demonstration 3* shows the effect of distance and angle on the overlap between *p*-orbitals.

Demo 3 Interactive orbital overlap diagram between two *p*-orbitals; MOs and bond energy as a function of bond distance and angle.

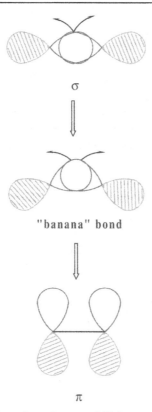

σ

"banana" bond

π

Fig. 2.8 The transformation of a σ- into a π-MO by rotation of the two *p*-orbitals away from each other. The intermediate situation is a bent, or "banana" bond, as found in strained compounds

2.4
Combining AOs to Build MOs

Molecular orbitals, which we will also later use as *group orbitals*, can be built from AOs in exactly the same way as MO-programs do, except that we can use the LCAO principle qualitatively to understand the AO-combination process. We will consider a simple example, methylene, CH_2, in order to illustrate the principles involved. We can then use the MOs obtained as generic orbitals for the fragment or group AH_2, where A can be any main group element, in order to explain the shapes of these molecules, and also as group orbitals in order to build the MOs of more complicated molecules like ethylene or cyclopropane.

2.4.1
Methylene, CH₂

The MOs of methylene are built from the AOs of one carbon and two hydrogens within the LCAO approximation. However, because the two hydrogens are symmetrically equivalent, their AOs cannot be considered separately, but must be combined to *symmetry-adapted combinations*. This is because the CH_2 molecule has C_{2v}-symmetry, for which the most relevant *symmetry element* in this discussion is the *mirror plane* shown in Fig. 2.9. An introduction to symmetry elements will be given in Sect. 5.1.

The symmetry-adapted combinations of hydrogen s-orbitals that we use to build the methylene MOs must be either symmetric or antisymmetric with respect to reflection in this plane. The individual AOs do not fulfill this condition, but can be combined to give the two symmetry-adapted combinations shown in Fig. 2.9. These combinations can then be used to build the MOs.

Once we have built symmetry-adapted combinations of AOs for all equivalent sets of atoms, we can begin to combine them to form the MOs

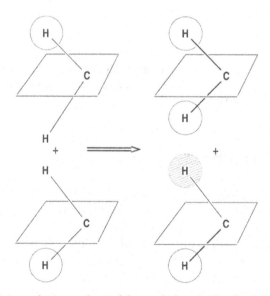

Fig. 2.9 The horizontal mirror plane of the methylene molecule makes it necessary to combine the two individual hydrogen AOs (left) to give the symmetry-adapted combinations shown on the right

of methylene. Symmetry is a great help in this process, allowing us to determine which AOs and symmetry-adapted combinations can interact with each other and which not. Orbitals that cannot interact with each other are said to be *orthogonal*. The horizontal mirror plane already used to obtain the symmetry-adapted combinations of hydrogen AOs and the one situated in the molecular plane suffice to distinguish all the different symmetry types involved in forming the methylene MOs from the AOs used here. Figure 2.10 shows the two different symmetry planes that we will use.

We can now place the two symmetry-adapted combinations of hydrogen s-orbitals and the four carbon AOs at their correct positions in methylene and classify them as to whether they are symmetric or antisymmetric with respect to each of the two mirror planes. This is shown in Fig. 2.11.

Only one orbital, the carbon p_x, is antisymmetric with respect to reflection in the σ_{yz}-plane. This AO is thus orthogonal to all the others and will be used unchanged in the methylene MOs. The carbon p_y-orbital is antisymmetric with respect to reflection in the σ_{xy}-plane, as is the symmetry-adapted combination denoted ψ'_{HH}. These two orbitals can thus interact with each other, as shown in Fig. 2.12.

The resulting two MOs are bonding (π_{CH2}) and antibonding (π^*_{CH2}) combinations of the carbon p-AO and the antisymmetric combination of hydrogen s-AOs. We will discuss the details of these MOs in more detail when we consider the quantitattive aspects of bonding in methylene, but note that the carbon contribution is purely p, making it less favourable than an MO in which carbon uses its s-orbital. These two MOs are often denoted π because they are antisymmetric with respect to their nodal plane. This designation does not imply that the MO is part of a conjugated π-system.

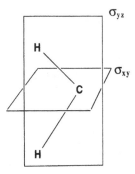

Fig. 2.10 The two mirror planes used to characterise the symmetry of orbitals for methylene

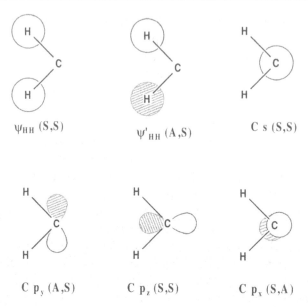

Fig. 2.11 The carbon AOs and hydrogen symmetry-adapted combinations of AOs used to build the MOs of methylene. The letters in parentheses denote symmetric (S) or antisymmetric (A) with respect to the σ_{xy} and σ_{yz} mirror planes, in that order

The shape of the orbitals does, however, let them interact very effectively with π-systems, so that they are important in *hyperconjugation*.

The three remaining orbitals, the symmetrical ψ_{HH}, the carbon s and p_z, are all symmetrical with respect to both mirror planes. They can therefore all interact with each other. The resulting MOs are shown in Fig. 2.13.

Figure 2.14 shows the electron density due to each of the six MOs of methylene with their energy levels calculated at the AM1 semiempirical level of theory. Note that the mixing of the s- and p_z-AOs on carbon leads to the familiar orbital shapes described by hybrid orbitals in valence bond theory. The σ_{CH}-orbital is composed almost exclusively of s-AOs, so that the negative lobe remains very small. The higher orbitals have more p-character on carbon.

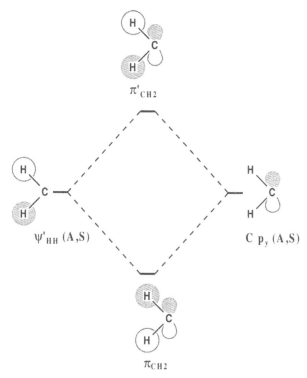

π'_{CH2}

ψ'_{HH} (A,S)

C p_y (A,S)

π_{CH2}

Fig. 2.12 The interaction of the carbon p_y-AO with the antisymmetric combination of hydrogen s-orbitals to give the CH-bonding π_{CH2}-MO and the antibonding π^*_{CH2}-MO. These MOs are designated π because of their nodal plane. They are not part of a conjugated π-system, but can be involved in hyperconjugation

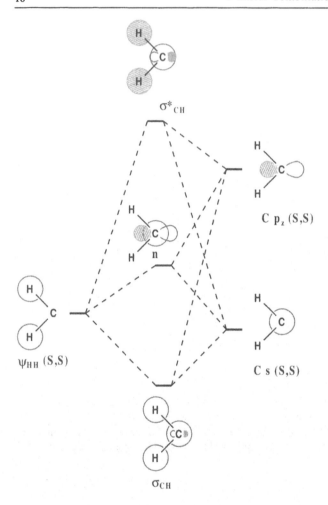

Fig. 2.13 The interaction between the three symmetric MOs used to build the MOs of methylene. The σ_{CH}-MO is CH-bonding and consists mainly of σ-contributions and is therefore particularly stable. The n-MO is a carbon lone-pair (and therefore essentially non-bonding). The σ^*_{CH}-MO is the antibonding counterpart of the σ_{CH}-MO

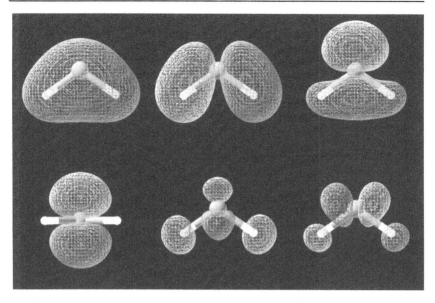

Fig. 2.14 AM1-calculated electron densities and energy levels for the valence molecular orbitals of singlet methylene

2.5
The AH₂ Walsh Diagram

As mentioned above, the MOs derived in Sect. 2.4 for methylene are applicable to any AH_2 fragment where A is a main group element. So-called *Walsh diagrams* consider the changes in the shapes and energies of individual MOs as the geometry is changed, in this case the dependency of the MOs of AH_2 fragments on the central H-A-H angle. Figure 2.15 shows the Walsh diagram that allows us to estimate the central angle in AH_2.

 Interactive Walsh diagrams for AH_2, AH_3 and AH_4; orbitals and energy levels as a function of the geometry.

The σ_{CH}-orbital consists essentially of an *s*-orbital on carbon. *s*-Orbitals are spherical and therefore overlap equally well with ligand orbitals in any direction. The energy of the σ_{CH} is therefore largely unaffected by the change from a bent to a linear geometry. The π_{CH2}-MO, on the other hand,

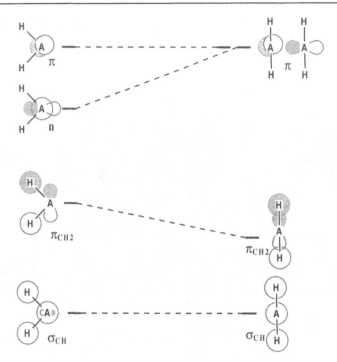

Fig. 2.15 Walsh diagram for AH_2 molecules with a bent (left) and a linear (right) H-A-H angle

is quite sensitive to the H-A-H angle. As shown in Sect. 2.3, overlap with p-orbitals is at a maximum when the ligand lies on the principal axis of the p-AO. By moving the hydrogen ligands from a bent to a linear geometry, we move them into the principal axis of the carbon-p, and therefore to the position of maximum overlap. The π_{CH2} is therefore more stable in the linear geometry than in the bent. The energy of the n- (lone pair) orbital depends strongly on the H-A-H angle for another reason. In the bent geometry, the s- and p-AOs on carbon can mix with each other. This mixing can be considered to stabilise the p-orbital. At the linear geometry, however, the s- and p-AOs are orthogonal, so that the lone pair becomes a pure p-orbital that is *degenerate* with the original π-MO, which remains unaffected by the change in geometry because it is localised on the carbon. Thus, the energy of the n-orbital rises steeply on changing the geometry from bent to linear.

These changes in orbital energy can be used to rationalise the observed geometries of AH_2 fragments. Figure 2.16 shows the energy levels given in

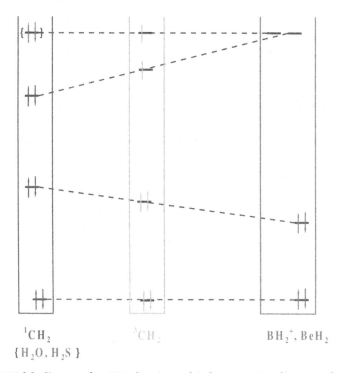

Fig. 2.16 Walsh diagram for AH$_2$ showing orbital occupation diagrams for 4-, 8- and singlet and triplet 6-electron species

Fig. 2.15 with schematic orbital occupations for singlet and triplet CH$_2$, H$_2$O and H$_2$S, BH$_2^+$ and BeH$_2$.

Singlet methylene, the molecule used to derive the MOs in Sect. 2.4.1, has six valence electrons. These occupy the lowest three MOs, the σ_{CH}, the π_{CH2} and the n. As described above, the σ_{CH} has no preference for a bent or linear geometry, whereas the π_{CH2} prefers linearity and the n-lone pair a bent geometry. The latter two orbitals therefore oppose each other in their geometric preferences and we must decide which of the two has the stronger preference (i.e. which slope of the two orbital correlation lines is larger). In this case, the n-lone pair dominates and singlet methylene is strongly bent. Adding two electrons to obtain the 8-electron species water or H$_2$S has little or no effect on the preferred H-A-H angle because the two additional electrons occupy the π-MO, which is independent of the angle.

Triplet methylene has a larger H-C-H angle than the singlet because the n-orbital is only singly occupied and therefore half as important energetically as in the singlet. The geometric preference of the π_{CH2} is therefore more important for the triplet and the bond angle is larger. Finally, the four-electron species BH_2^+ and BeH_2 have only the σ_{CH} and π_{CH2} orbitals occupied and therefore prefer linear geometries. Table 2.1 gives the bond angles found for the species shown in Fig. 2.16.

Note that H_2S has a smaller bond angle than $^1CH_2$ or H_2O, although they are all have predicted to have similar geometries by the Walsh diagram. This is a general trend between molecules involving main group elements of the first and higher long periods.

Walsh diagrams are a powerful tool for explaining the geometries of covalent molecular fragments, but can fail for highly ionic species. One well know example is Li_2O, which ought to have a geometry similar to that of water according to the Walsh approach, but is in fact found to be linear (Fig. 2.17). The reason for this discrepancy is that the bonding in Li_2O is not described well by a covalent treatment such as that implicit in the Walsh diagram shown above and that the bond angle is determined by the electrostatic repulsion between the two positively charged lithium ions.

Table 2.1 Bond angles found for the species shown in Fig. 2.16

Species	$^1CH_2$	H_2O	H_2S	$^3CH_2$	BH_2^+	BeH_2
Angle (°)	101.9	103.5	92.1	132.6	180.0	180.0

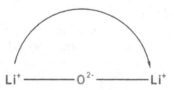

Fig. 2.17 The 180° bond angle in Li_2O minimises the electrostatic repulsion

3 Larger Molecules

The group orbitals introduced in Chap. 2 and used to interpret structures using Walsh diagrams also form an ideal basis for building up and understanding the MOs of large molecules. This will be demonstrated for several examples below, but before discussing the examples in detail, we should consider the two alternative approaches to understanding molecular structures and reactivity, molecular orbital- and *valence bond- (VB) theory.*

The MO- and VB- approaches are *alternative* ways to look at the same problem – chemical bonding. This is unfortunately not always made completely clear in organic or inorganic textbooks, but rather the two approaches are used side by side without any consideration of their relationship to each other. Often, chemists use the treatment that is best suited for the problem in hand. It is, for instance, very difficult to find a molecular orbital description of an aldol condensation in current textbooks, whereas the stereoselectivity of *electrocyclic reactions* is usually discussed using MO-theory. This is because an aldol condensation, which can be treated using MO-theory, can be described much more succinctly using typical "arrow pushing" (i.e. VB) arguments (Fig. 3.1).

An electrocyclic reaction, on the other hand, can be depicted using a VB-picture, but its stereochemistry cannot be deduced from such a treatment (Fig. 3.2).

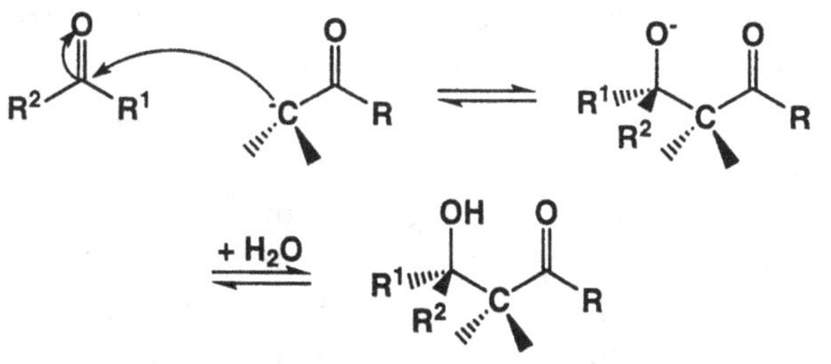

Fig. 3.1 Schematic representation of the aldol condensation

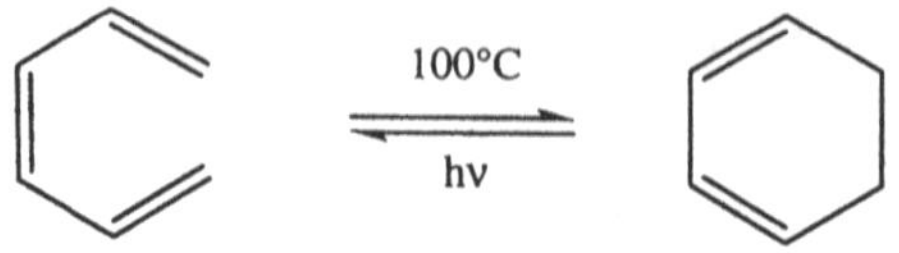

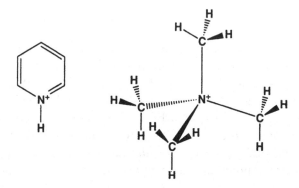

Fig. 3.2 Electrocyclic interconversion of 1,3,5-hexatriene to 1,3-cyclohexadiene

Fig. 3.3 Pyridinium and tetramethylammonium cations

Generally, a VB-treatment is used to teach chemistry (because it is more succinct and less complicated than MO-theory) until cases (like electrocyclic reactions) are treated that cannot be understood properly within simple VB-theory. This need not be the case, but is usually a good compromise for teaching chemical bonding and reactivity effectively.

There is, however, one quantity, atomic charge, that has very different meanings within the two theories. Simple VB-theory uses *Lewis structures* to describe the bonding in molecules. Formal Lewis charges are assigned in order to complete the octet of a given atom, but do not indicate that the atom in question actually has the formal charge (atomic charges are not physically measurable quantities, but can be assigned by a variety of analyses within MO-theory). Take, for instance, the pyridinium and tetramethylammonium cations (Fig. 3.3).

The nitrogen is the most electronegative atom in these ions, so that it is the last place that we would expect to find a positive charge. In fact, whichever method we use to calculate the atomic charges, the nitrogens turn out to be slightly negative. The positive charge in the tetramethylammonium ions is almost completely situated on the 12 hydrogens.

We will now treat some common molecules using qualitative MO-theory. It is, however, important to remember that this treatment is an alternative to the common Lewis (VB) picture. We will point out some of the differences for the individual molecules.

3.1
Ethylene

We could start from the AOs of two carbons and four hydrogens to derive the MOs of ethylene (which is exactly what MO-programs do), but it is far more effective and easy to understand if we start from the group orbitals of two methylenes. These are exactly the orbitals derived in Sect. 2.4.1, but we will limit ourselves to the two CH-bonding, the n- and the π-orbitals. The two remaining antibonding orbitals can be treated exactly analogously. The orbital-combination-procedure is very simple. We simply have to form bonding and antibonding combinations of each of the different types of group orbital (see Fig. 3.4), starting from the lowest, the σ_{CH}.

Because the electron density of these two orbitals is not directed along what will be the C = C bond, their overlap is relatively small and the energy splitting between the bonding and antibonding combinations is also small (Note that we use "bonding" and "antibonding" in this context to denote the relative phases of the two group orbitals, not that a new bonding interaction is formed.)

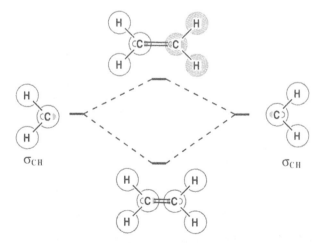

Fig. 3.4 Formation of the lowest lying σ_{CH}-orbitals of ethylene

We can use exactly the same technique to combine the second CH-bonding group orbitals, the π'_{CH2} (see Fig. 3.5).

The above two combinations demonstrate one of the guiding principles of understanding molecular orbitals: MOs can usually be considered to be either bonding or antibonding combinations of group orbitals. In the above two cases, no new bond is formed (we take four CH-bonding AOs and form four CH-bonding MOs), and thus the bonding and antibonding combinations are not split as strongly as they would be if a new bond were formed, which is the case for the n- and π-group orbitals, as demonstrated in Figs. 3.6 and 3.7.

Note that the splitting between the σ- and σ^*- MOs is considerably larger than found for the CH-bonding group orbitals and that a new σ-bonding MO is formed. This large splitting will play a significant role in determining the energetic ordering of the ethylene MOs.

The final group orbitals that we will consider here are the π-orbitals prependicular to the CH_2-plane. These MOs are formed as combinations of the pure carbon p-orbitals that form the *Lowest Unoccupied Molecular Orbital (LUMO)* of singlet methylene and the highest *Singly Occupied Molecular Orbital (SOMO)* of the triplet.

The valence electrons (4 each for the two carbons and 1 each for the hydrogens = 12) can now be assigned to the MOs according to their ener-

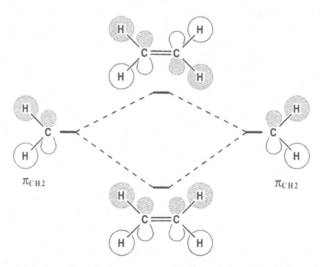

Fig. 3.5 Combination of methylene π_{CH2}-orbitals to form two ethylene CH-bonding MOs

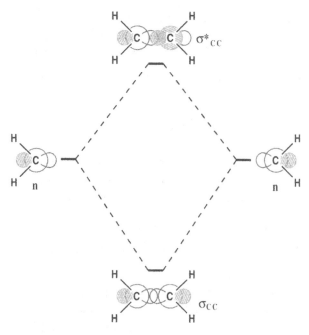

Fig. 3.6 Combination of methylene n-orbitals to form the CC-bonding and anti-bonding σ-orbitals of ethylene

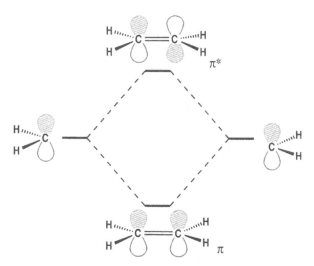

Fig. 3.7 Formation of the π and π^*-MOs of ethylene

getic ranking. The orbital energies are not given by the qualitative group-orbital-combination-technique outlined above, but generally the correct energetic ordering can be deduced using the following simple rules:

1. Bonding orbitals are more stable than nonbonding, which in turn are more stable than antibonding.

If the characters (bonding, nonbonding, antibonding) of two orbitals are the same:

2. σ-MOs are more stable than π, π^* are more stable than σ^*.
3. Orbitals with high coefficients on electronegative atoms are more stable than those that are concentrated on more electropositive ones.

If the central elements of the MOs are the same:

4. MOs with high s-coefficients are more stable than those that are mainly p.

This gives the following orbital scheme for ethylene (Fig. 3.8).

Note that in this scheme only the π^* unoccupied or *virtual orbital* has been included. There are, of course, unoccupied antibonding equivalents for all 6 occupied valence MOs (the π^*, and, not shown, the σ^*_{CC} and the four CH-antibonding MOs derived from the CH-antibonding group orbitals).

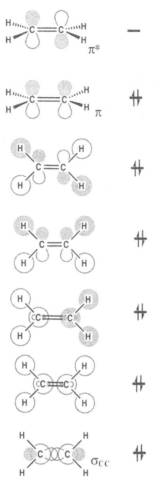

Fig. 3.8 Schematic diagram of the occupied valence MOs and LUMO of ethylene

3.2
Cyclopropane

The molecular orbitals of cyclopropane can be built from the CH_2 group-orbitals in exactly the same way as shown above for ethylene. We will, however, limit the discussion of cyclopropane to the CC-bonding and antibonding MOs, although the complete set of CH-bonding and antibonding MOs can be derived in the same way from the corresponding methylene group-orbitals. For the CC-bonding and antibonding MOs we need only the methylene n- and π-orbitals given in Fig. 3.9.

If we first consider the n-orbitals, we can form three combinations from the three methylene groups (see Fig. 3.10).

The most stable of these is that in which all three orbitals enjoy bonding overlap to form a completely bonding, symmetrical combination. This MO will clearly be very stable because it is strongly bonding and because it consists of a very high proportion of contributions from s-AOs. The remaining two MOs are energetically degenerate, and can only be constructed with a sum of one antibonding interaction. *Degenerate orbitals will be treated in Sect. 5.4*, so it suffices to note here that the degenerate set

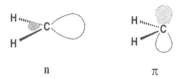

n   π

Fig. 3.9 The n- and π-group orbitals of methylene

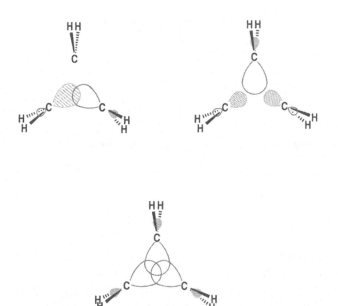

Fig. 3.10 Combination of three methylene n-group orbitals in cyclopropane

formed by combination of three methylene n-orbitals is drawn as if the symmetry of the molecule had been reduced to make them slightly different in energy and either symmetric or antisymmetric relative to a mirror plane perpendicular to the ring plane and including the uppermost methylene group. Quite generally, combination of three equivalent AOs or group orbitals in a symmetric geometry leads to a single MO and a degenerate pair. This can also be seen, for instance, for the cyclopropenium cation or for the π_{CH2}-group orbitals for CH_3. The important feature of the degenerate σ_{CC}-orbitals in cyclopropane is that they are net antibonding, and therefore lie high in energy (Fig. 3.11). Thus, only one bonding combination can be formed from the energetically relatively low lying n-group orbital of methylene. This clearly represents a significant energetic disadvantage for cyclopropane relative to an acyclic alkane, which would be expected to be able to use the methylene n-group orbitals better. This factor alone should result in *strain energy* in cyclopropane. In molecular orbital terms, strain results when the molecule cannot form MOs that are as stable as those found in the strain-free reference molecule(s). These orbitals are, however, not the main cause of strain in cyclopropane. Not only must the remaining two σ_{CC}-orbitals be built up from the energetically unfavourable π-group orbitals of methylene (remember this orbital is the LUMO in the singlet carbene), but they are also unfavourable because there are no com-

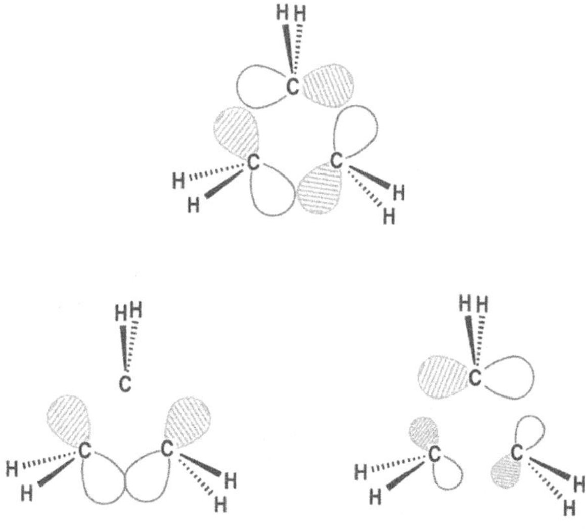

Fig. 3.11 Combination of three methylene π-group orbitals in cyclopropane

pletely bonding combinations. In this case, there are two degenerate bonding and a single antibonding MOs.

The two degenerate orbitals are net bonding and are the HOMOs in cyclopropane. The nature of these two orbitals was first pointed out by Walsh, and they are therefore often called *Walsh orbitals*. The Walsh orbitals are particularly unstable relative to other σ_{CC}-MOs both because they are made up exclusively of *p*-contributions and because the bonding overlap is less efficient than in unstrained CC-bonds. In order to understand this, we should consider the overlap between two *p*-orbitals in different orientations to each other (Fig. 3.12, see also Sect. 2.3).

The overlap between two *p*-orbitals is greatest when the axes of the two orbitals are collinear, as is the case for the σ-bond shown above. The minimum overlap (at a fixed interatomic distance) is found when the two orbital axes are parallel to each other, as found in π-bonds. The situation for the Walsh orbitals in cyclopropane is intermediate between these two extremes. Thus, Walsh orbitals represent a situation in which much of the bonding electron density is situated outside the line joining the two atoms. Such bonds are often called "bent" or "banana" bonds. They lie higher in energy than normal σ-bonds and therefore are a manifestation of strain in the molecule. They are, however, also responsible for the unusual reactivity of cyclopropanes, which undergo some reactions, such as hydrogenation or bromi-

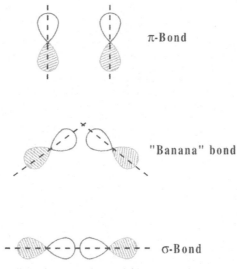

Fig. 3.12 σ-, "banana", and π-MOs formed from two *p*-orbitals

nation, that are more typical for olefins than for alkanes. These reactions lead to ring-opening in cyclopropanes and addition to the double bond in alkenes.

3.3
π-Systems

Much of the pioneer work using MO-theory was limited to π-systems. One reason was technical – early MO-techniques only treated the π-system and ignored σ-bonds – and another was the fact that many chemical reactions occur by reaction with one or more double bonds. Figure 3.13 shows typical energy levels for the different types of molecular orbitals.

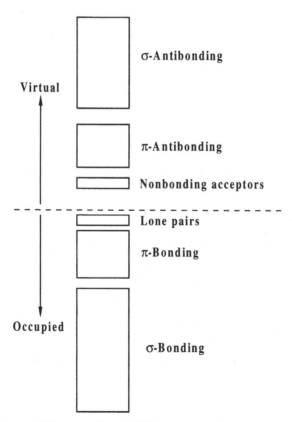

Fig. 3.13 Schematic diagram of typical MO energy levels

The most stable orbitals are generally σ-bonds, whereas π-bonding orbitals lie higher in energy because of the overlap effect described in the previous section. Note that this energetic sequence also explains the reactivity of π-systems. *Frontier orbital theory* (see Sect. 4.2) suggests that MOs close to the HOMO-LUMO border are more important in determining reactivity than more stable occupied or less stable virtual orbitals. Nonbonding lone pairs generally lie higher in energy than MOs from the π-system. The order is reversed for the unoccupied, or virtual orbitals. Here nonbonding acceptors, such as those found for carbocations or *Lewis acids* in general, are lowest in energy (often with negative energies, which means that the system can accept an electron with a gain in energy according to *Koopmans' theorem*). The antibonding orbitals of the π-system come next, followed by the σ-antibonds, which are often very high in energy. Note that this orbital ordering, and the *p*-overlap arguments given in Sect. 3.2, suggest that π-bonds can be regarded as very strained and reactive σ-bonds. One consequence of this is that electrocyclic reactions usually occur in the direction in which two π-bonds are converted into two new σ-bonds.

In the following, we will describe two simple techniques for determining the character of π-orbitals in linear and cyclic systems.

3.3.1
Linear Systems

The character of the π-MOs in a linear conjugated system can be determined using three simple rules:

1. The lowest energy orbital is bonding throughout, and therefore symmetrical relative to a central nodal plane.
2. The symmetry of the MOs relative to a central nodal plane alternates as the energy increases
3. The number of nodal planes between *p*-orbitals (not the plane of the π-system itself) increases by one for each MO as the energy increases.

This principle is illustrated for 1,3,5-hexatriene in Fig. 3.14.

Such a simple scheme does not give the magnitudes of the AO-coefficients, but is useful simply to determine the nature and symmetry of the individual MOs.

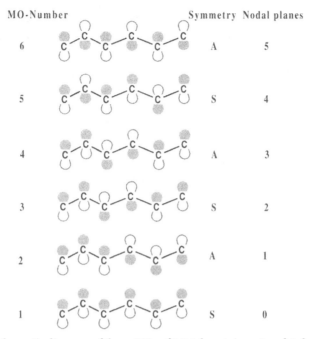

Fig. 3.14 Schematic diagram of the π-MOs of 1,3,5-hexatriene. S and A denote symmetry and antisymmetry relative to a central nodal plane

3.3.2
Cyclic Systems

Similar principles apply to cyclic π-systems, with the slight modification that degenerate orbitals can occur and that there is a very simple technique, proposed by Frost and Martin, to determine the MO-pattern and *Hückel theory* energy levels. These can be obtained simply by drawing a circle and then placing a regular polygon with the correct number of sides with one corner at 6 o'clock within it. The energy levels are then given by the positions of the corners, as illustrated in Fig. 3.15 for benzene.

We thus find the familiar pattern of a single low lying MO followed by two doubly degenerate sets and a final single high energy orbital.

The nature of the MOs can now be determined qualitatively using the fact that the number of nodal planes starts at zero for the lowest orbital and increases by one for each energy level, giving the well-known benzene

MOs (note that the degenerate sets are portrayed as symmetric and anti-symmetric relative to a vertical plane perpendicular to the page (Fig. 3.16).

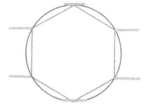

Fig. 3.15 Frost and Martin's technique illustrated for benzene

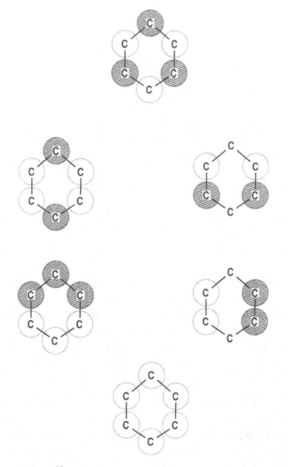

Fig. 3.16 The π-MOs of benzene

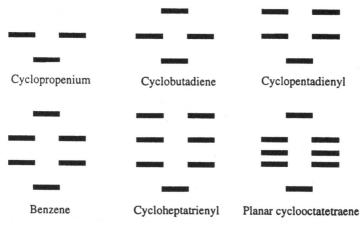

Fig. 3.17 Energy levels for simple annulenes

The π-MO patterns for the common cyclic conjugated systems can all be determined in this way. The resulting MOs are given in Chap. 6 and the patterns of the energy levels in the Fig. 3.17.

3.4
Hyperconjugation

Hyperconjugation is defined as the (usually stabilizing) interaction of an occupied σ-bonding orbital with a vacant orbital (usually a p- or π-orbital). *Negative hyperconjugation*, which is far less common and whose existence was controversial for many years, is the interaction of a high lying occupied orbital (usually of an anionic centre) with a vacant σ*-antibonding orbital. Thus, hyperconjugation is a major stabilising factor for carbocations and negative hyperconjugaion occurs for carbanions. Negative hyperconjugation is closely related to the *anomeric effect*, which will also be discussed below.

3.4.1
The Ethyl Cation

The ethyl cation is the prototype system for demonstrating the effect of hyperconjugation. Consider classical $CH_3CH_2^+$ as a combination of a methyl group and a $-CH_2^+$ centre. The group orbitals of the methyl group (equivalent to the MOs of NH_3) include the π_{CH3}-orbital shown schemat-

ically in Fig. 3.18a. This orbital has the correct nodal characteristics to interact strongly with the unoccupied p-orbital of the $-CH_2^+$ group (shown in
Fig. 3.18b).

The "correct nodal characteristics" in this case means that both orbitals
have a nodal plane containing the connecting bond and perpendicular to
the page in Fig. 3.18. This means that they can overlap strongly with each
other to give the interaction diagram shown in Fig. 3.19.

Note that the two orbitals have the same symmetry, which is a necce-
sary condition for non-zero overlap, but that this is not sufficient for strong
overlap. In this case, the only symmetry element is a mirror plane corresponding to the plane of the page, but the fact that both orbitals have a nodal
plane as described ensures that they can overlap strongly. This sort of consideration, which involves what might be called "approximate symmetry"
is common in MO-arguments and will be considered again for the Woodward-Hoffman rules. The two orbitals form bonding and antibonding combinations with contributions that depend on the relative stabilities of the
two component group orbitals, exactly analogously to the interaction diagram shown for LiH in Fig. 2.5. Because only the bonding combination is
occupied, this interaction results in a stabilisation. This is usually expressed
as a stabilisation of the cationic centre by the occupied orbital but, because
the orbital involved on the cationic centre was originally vacant, it is more
accurate to say that the σ-system is stabilised by interaction with the cation
centre. This interaction can be seen in the HOMO-2 and the LUMO of the
staggered ethyl cation, which correspond to the bonding and antibonding
combinations of group orbitals shown in Fig. 3.19.

One consequence of the hyperconjugative overlap shown above is that
electron density is removed from the CH-bond of the methyl group (and

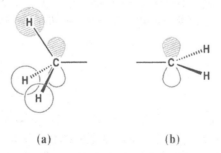

(a)   (b)

Fig. 3.18 The π_{CH3} orbital of the methyl group and the unoccupied p-orbital of the
cationic centre involved in hyperconjugtion in the ethyl cation

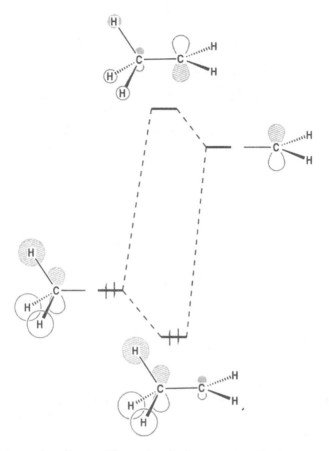

Fig. 3.19 Interaction diagram illustrating the hyperconjugation interaction in the classical ethyl cation

especially the one in the plane of the page) and that bonding overlap occurs between the methyl hydrogens and the cationic carbon. In most cases, this results in a shortening of the connecting bond and a lengthening of the hyperconjugating bonds. In the ethyl cation, however, the interaction is so strong that it leads to a symmetrically bridged structure in which the unique hydrogen of the methyl group is shared between the two carbons. The two orbitals shown above then become the HOMO-1 and the LUMO of the bridged ethyl cation, which is the *global minimum* structure for this cation (Fig. 3.20).

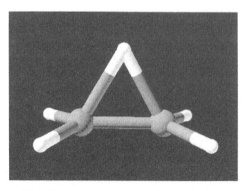

Fig. 3.20 Structure of the bridged "non-classical" ethyl cation

Hyperconjugation is the reason that highly substituted cations such as
^tbutyl are very stable. Note also that carbon-carbon bonds generally hyper-
conjugate better than carbon-hydrogen bonds.

3.4.2
The Cyclopropylcarbinyl Cation

It can be seen from Fig. 3.19 that the smaller the energy gap between the
occupied σ-MO and the unoccupied orbital on the cationic centre, the larger will be the stabilisation. One way to obtain high energy σ-MOs is to consider strained molecules such as cyclopropane. One of the two degenerate
Walsh orbitals of cyclopropane shown in Fig. 3.11 has the correct nodal
characteristics to interact with a cationic centre in the correct orientation.
As this orbital lies particularly high in energy, it should be an especially efficient hyperconjugator. The two interacting orbitals are shown in Fig. 3.21.

This interaction can be seen particularly well in the HOMO-1 of the
bisected cyclopropylcarbinyl cation, but also to some extent in the LUMO.
Once again, the hyperconjugation is so strong that the cation distorts to a
non-classical structure that is not reproduced by the AM1 calculations used
in this book. Another consequence of the hyperconjugation is the large calculated energy difference (14 kcal mol^{-1} at AM1) between the bisected and
perpendicular conformations of the cation, even though the perpendicular conformation also enjoys some hyperconjugation, as shown by the
HOMO-5 and the LUMO (which cyclopropane MO is involved?).

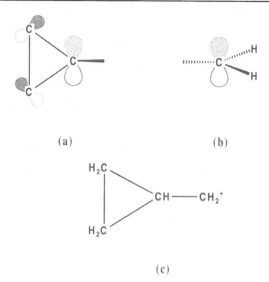

(a)   (b)

(c)

Fig. 3.21 The high lying Walsh orbital of cyclopropane (a) and the accepting orbital of the cationic centre (b) involved in the strong hyperconjugative stabilisation of the cyclopropylcarbinyl cation (c)

3.4.3
Negative Hyperconjugation

Just as the acceptor orbitals at cationic centres can interact with high lying σ-MOs, the high energy donor orbitals at anionic centres can interact with low lying σ*-orbitals. This effect is known as negative hyperconjugation. It occurs when an electronegative atom, X, such as oxygen or fluorine is bound to a carbon atom attached to an anionic centre. The CX-bond is strongly polarised towards the electronegative element, so that the σ^*_{CX}-MO is therefore concentrated on carbon, analogously to the LiH MOs shown in Fig. 2.5. This polarisation towards carbon allows a strong overlap with the non-bonding electron pair at the anionic centre, as shown in Fig. 3.22.

This interaction leads to a stabilisation of the anionic centre, but also to a weakening of the C-X-bond and partial double bond character in the CC-bond. If these changes are extrapolated further (Fig. 3.23), they lead to elimination of X⁻ to give an olefin.

The HOMO and the LUMO+2 of the σ-fluoroethyl anion demonstrate negative hyperconjugation.

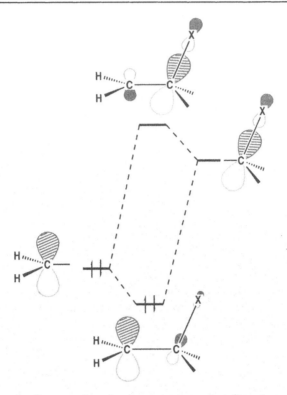

Fig. 3.22 Negative hyperconjugation between a non-bonding electron pair at an anionic centre and a low lying σ_{CX}-orbital, where X is an electronegative element

Fig. 3.23 VB-approach to negative hyperconjugation

A further impressive example of negative hyperconjugation, this time in a neutral compound, is NF_3O, whose highest occupied and lowest unoccupied MOs show strong mixing between lone pairs and σ^*-orbitals. This leads to unusual stability and bond lengths.

3.4.4
The Anomeric Effect

The anomeric effect is simply negative hyperconjugation under another name. It is best known as the effect that stabilises the normally less stable axial position of oxygen substituents on sugars and can be demonstrated using the axial and equatorial hydroxypyranes in Fig. 3.24.

In the axial conformation, negative hyperconjugation between the higher lying of the two lone pairs on the ring oxygen and the exocyclic σ^*_{CO}-orbital leads to a stabilisation that is not possible in the equatorial conformation, as shown on Fig. 3.25.

In the alternative equatorial conformation, only a much weaker interaction between the σ^*_{CO}-orbital and the lower of the two ring oxygen lone pairs is possible. This leads to the unusual preference of the hydroxyl substituent for the axial position. The AM1 calculations predict that the axial conformation is 0.7 kcal mol^{-1} more stable than the equatorial that is also found in sugars. Thus, the anomeric effect is simply another manifestation of negative hyperconjugation, using an oxygen lone pair instead of an anionic centre. These effects can be seen in the HOMO and LUMO+1 of

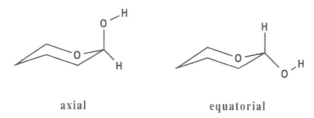

axial   equatorial

Fig. 3.24 Axial (left) and equatorial (right) conformers of 2-hydroxypyrane

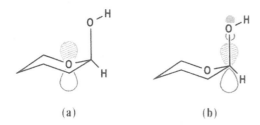

(a)   (b)

Fig. 3.25 The highest lying ring oxygen lone pair (a) and the exocyclic σ^*_{CO}-orbital (b) responsible for the anomeric effect in the axial conformation

the axial 2-hydroxypyrane. It is a useful exercise to identify the orbitals involved in the alternative interaction in the equatorial 2-hydroxypyrane from the HOMO and LUMO+1. The anomeric effect is also responsible for the fact that polyhalogenated methanes have increasing CX-bond strengths with increasing degree of halogen substitution because the lone pairs of the halogens interact with adjacent σ^*_{CX}-orbitals (analogously to NF_3O).

4 Reactions

4.1
Lewis Acid/Lewis Base Interactions

Lewis acids are electron acceptors and *Lewis bases* electron donors. This means that the former, as we can see using BH_3 as an example, have at least one low-lying unoccupied orbital. In the case of BH_3, this is the LUMO, a pure boron p-orbital. The AM1-calculated energy for this MO is +1.6 eV, a low value for the LUMO of a neutral compound. This means that BH_3 can accept an extra electron to form the BH_3 radical anion. When the extra electron is added, the BH_3 moiety becomes pyramidal, as shown in Sect. 2.5 (Walsh diagrams). We can now use ammonia as an example of a Lewis base. The lone pair HOMO has a calculated energy of –10.4 eV (i.e. a Koopmans' theorem ionization potential of 10.4 eV). We can remove an electron from this MO to form the ammonia radical cation, which has a planar trigonal structure.

If we now allow BH_3 and ammonia to interact with each other, we can expect the simple two-electron interaction shown in Fig. 4.1.

This interaction can be seen in the structure and MOs of the ammonia:borane complex. The donor- and acceptor-MOs interact to form a new bonding σ_{BN}- and antibonding σ^*_{BN}-orbitals. The σ_{BN}-MO is doubly occupied, resulting in a bonding interaction. Because, however, the ammonia HOMO was originally doubly occupied and the BH_3-LUMO empty, doubly occupying the shared σ_{BN}-orbital reults in transfer of negative charge from nitrogen to boron. The AM1-calculated charge on the NH_3-moiety in the complex is +0.48 (and consequently the charge on BH_3 –0.48) and the calculated dipole moment of the complex is 5.8 Debye with the positive end at nitrogen and the negative at boron. Thus, almost half an electron is transferred from the ammonia to the boron in forming the new bond. This sort of donor-acceptor (or Lewis base-Lewis acid) interaction is the basis

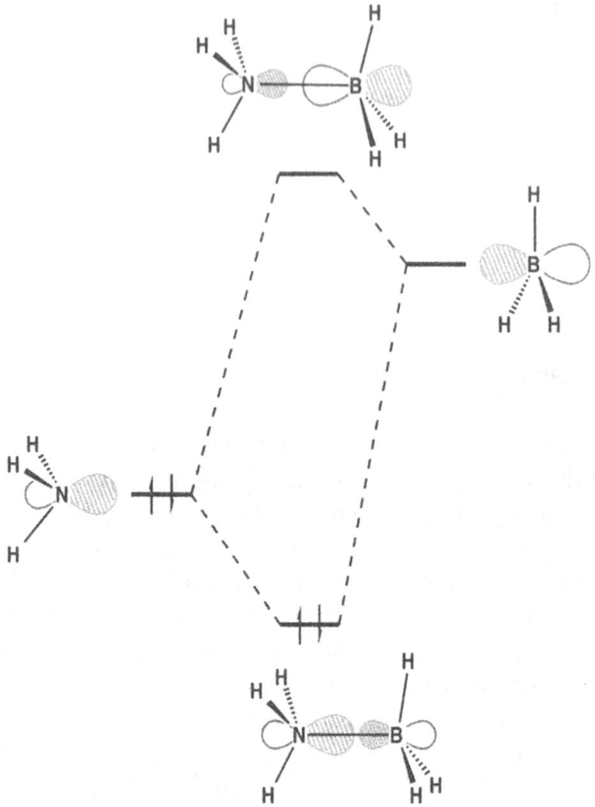

Fig. 4.1 The donor-acceptor interaction between the HOMO of ammonia and the LUMO of borane

of all electrophile/nucleophile chemistry. In general, the strength of the donor-acceptor interaction depends on the overlap betwen the donor and acceptor orbitals and on the reciprocal of the energy difference between them (see Sect. 4.2 below).

4.1.1
S_N2 Reactions

Acceptor orbitals need not be non-bonding. They can also be low-lying antibonding orbitals. One example of a reaction in which this is the case is binuclear nucleopilic substitution, or the S_N2 reaction. We will consider the model reaction of the fluoride anion (the nucleophile, donor or Lewis base) with fluoromethane. The σ^*_{CF}-LUMO+3 of fluoromethane does not lie particularly low in energy (F$^-$ is not a very good leaving group), but, because of the electronegativity difference between carbon and fluorine, is concentrated on carbon with a very large lobe on the opposite side to the fluorine substituent. This orbital can enter into a donor-acceptor inter-action with one of the degenerate HOMOs of the fluoride ion to give a flu-oride:fluoromethane complex, as shown in Fig. 4.2.

This complex lies lower in energy than its constituents in the gas phase, but is not bound in solution. However, just as for the ammonia:borane com-plex, electrons are donated into the acceptor orbital, which is CF-anti-bonding. The net result is to form a new bond to the incoming fluoride and to weaken the existing CF-bond. At the S_N2 transition state for this reac-tion the two carbon-fluorine bond are equally long and correspond to roughly half a normal single bond, and the original σ^*_{CF}-orbital and the original fluoride lone pair are now the HOMO and HOMO-5 of the tran-sition state.

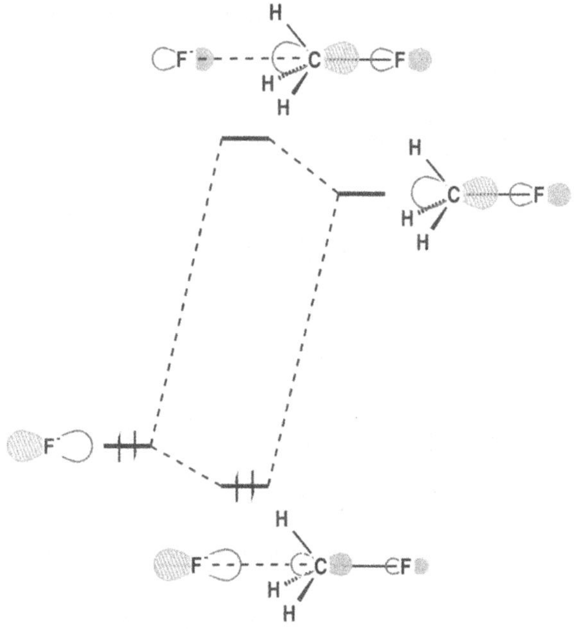

Fig. 4.2 Interaction of a fluoride lone pair with the σ^*_{CF}-orbital of fluoromethane

4.2
Selectivity; Frontier MO Theory

The strength of the interaction between a Lewis acid and a Lewis base can be expressed as the sum of the interactions between all the orbitals of the acid with all the orbitals of the base. However, the interaction energy between any pair of orbitals depends on the reciprocal of the energy difference between them. Furthermore, interactions between occupied and virtual orbitals are generally much larger than those between two occupied orbitals (although these can also be significant) and interactions between two virtual orbitals have no energetic consequences at all because there are no electrons involved. Therefore, the most important interactions energetically will be between the occupied and virtual orbitals with the smallest energy gaps. These are naturally the HOMO of the base and the LUMO of the acid. It is thus often a very useful approximation to use only the HOMOs and LUMOs (the frontier orbitals because they occur at the border between occupied and virtual orbitals) when considering reactivity. This approximation does not always work. It is, for instance, unreliable for extended π-systems with many high-lying occupied orbitals and low-lying virtual orbitals. It is, however, very useful for small, non-delocalized systems.

4.2.1
Nucleophilic Oxirane Ring-Opening

Let us now consider a special case of the S_N2 reaction, the nucleophilic ring-opening of substituted oxiranes. For 2,2-dimethyloxirane this reaction can occur in two ways (Fig. 4.3).

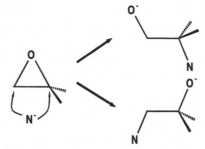

Fig. 4.3 Nucleophilic ring-opening of 2,2-dimethyloxirane

We saw above that nucleophilic substitution involves the σ^*_{CX}-orbital of the substrate; in this case a CO-antibonding orbital that is the LUMO of 2,2-dimethyloxirane. Visual inspection of this orbital shows that it has a larger contribution at the unsubstituted carbon atom than at the fully substituted one. This can also be seen from the atomic orbital coefficients in the MO. We can therefore conclude that, because the contribution at the unsubstituted carbon is larger, overlap with the HOMO of an attacking nucleophile N will be larger and the interaction energy will be more favourable. In this way it is possible to rationalise the experimental observation that under basic conditions 2,2-dimethyloxirane opens the ring to give the lower of the two products shown above. This sort of approach is often very useful in rationalising regioselectivity and sometimes stereoselectivity.

4.3
The Woodward-Hoffmann Rules

In order to understand the Woodward-Hoffmann rules for determining the stereochemistry of several different types of concerted reactions, let us first consider the ground and excited states of a "normal" reaction. The ground state energy rises continuously to the transition state and then falls to the product. Often, this is the only state shown in such reaction profile diagrams. The excited state is not generally involved in the reaction, but often has a minimum above the transition state, as shown in Fig. 4.4.

The reaction occurs in this case when the system gains enough energy to be able to cross the barrier to the product side. Let us, however, now con-

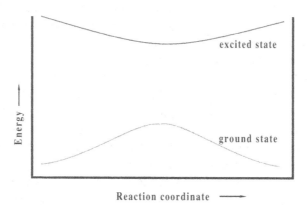

Fig. 4.4 Energy profiles for the ground and excited states of a "normal" reaction

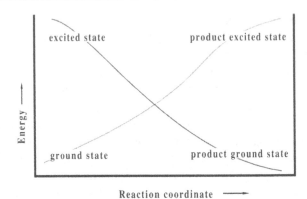

Fig. 4.5 Energy profiles for a reaction in which the ground state of the starting material does not lead to that of the product

sider a second reaction where the ground state energy profile does not lead directly to the product ground state, as shown in Fig. 4.5.

In this case, if the system stays in the same electronic state, it will gain energy continuously along the reaction path and lead to an excited state of the product. Similarly, if we were to start with the excited state of the reactant, the system could lose energy continuously to reach the ground state of the product. Strictly speaking, the two states can only cross if they have different symmetries, otherwise they will mix with each other near the crossing point to give a transition state for the ground state slightly below the state crossing. However, this activation energy is usually much higher in energy than those typically found for processes like that shown in Fig. 4.4. These reactions are known as *allowed reactions* and those as shown in Fig. 4.5 as *forbidden reactions*. The names are far stricter than the real situation; forbidden reactions can sometimes occur quite easily. However, the vast majority of reactions follow the allowed path if they have two alternatives.

In order to translate the above picture, which uses plots of the energies of different electronic states, into one that uses orbitals, we should consider a model case in which two electrons and only two orbitals are involved on either side of the reaction and in which we only consider the ground state and the doubly excited state in which both electrons occupy the higher orbital. This situation is shown in Fig. 4.6 for a forbidden reaction:

Because the dark and the light grey orbitals change their energy ordering, retaining their occupations along the reaction paths leads directly to

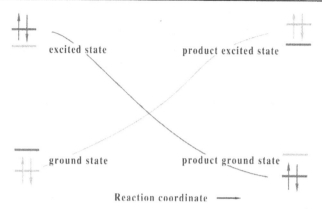

excited state   product excited state

ground state   product ground state

Reaction coordinate ⟶

Fig. 4.6 Orbital occupancies for a forbidden reaction as shown in Fig. 4.5

allowed   forbidden

Fig. 4.7 Orbital correlations for allowed and forbidden reactions

the excited states. This can be expressed for the ground state in terms of *orbital correlations*, which are shown in Fig. 4.7 for allowed and forbidden reactions:

Thus, the criterion for an allowed reaction in terms of state correlations diagrams like Figs. 4.4 and 4.5 is that the ground states of reactant and product correlate with each other. For orbital correlation diagrams like those shown in Fig. 4.7 the corresponding criterion is that no orbital correlation line crosses the border (the dashed line) between occupied and virtual orbitals. We will now consider three different sorts of reaction that can be treated in this way.

4.3.1
Electrocyclic Reactions

Electrocyclic reactions are those in which a π-system with N π-bonds and M rings is converted into a new system with N-1 π-bonds and M+1 rings (or the reverse). Some common examples are shown in Fig. 4.8.

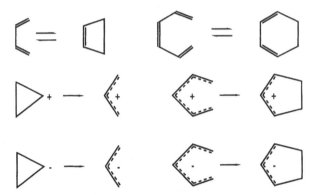

Fig. 4.8 Selected examples of electrocyclic reactions

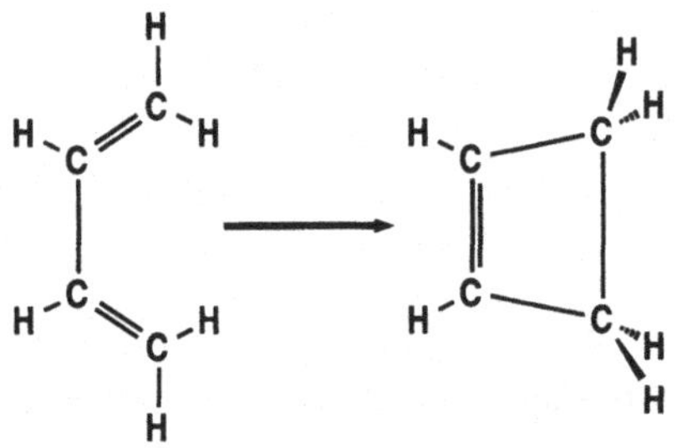

Fig. 4.9 Ring-closure of s-cis butadiene to cyclobutene

The net result of such electrocylclic reactions is the interconversion of a π- and a σ-bond. The reactions usually proceed towards formation of the σ-bond because σ-orbitals are lower in energy than π-, but formation of a strained ring may reverse this energetic effect. Let us now consider a prototype electrocyclic reaction, the ring-closure of *s-cis* butadiene to cyclobutene (Fig. 4.9).

This reaction can proceed in one of two ways. The two terminal CH_2-groups can either rotate in the same direction to give a *conrotatory* process (Fig. 4.10) or they can rotate in opposite directions to give the *disrotatory* ring closure (Fig. 4.11).

These two processes give the same product (cyclobutene) for the prototype reaction, but different stereoisomers if the butadiene is asymmetrically substituted with methyl groups. The purpose of the Woodward-Hoffmann rules for such reactions is to rationalise the stereochemistry of the products for both thermal and photochemical reactions.

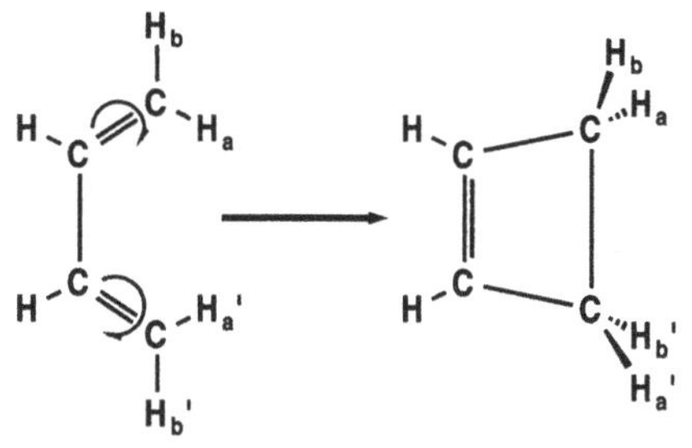

Fig. 4.10 *Conrotatory* ring-closure

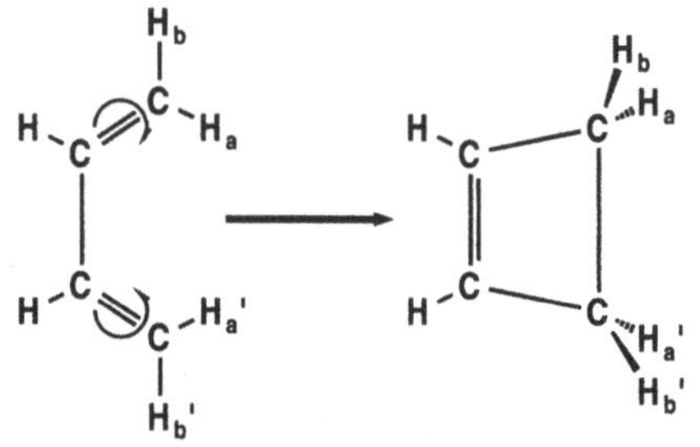

Fig. 4.11 *Disrotatory* ring-closure

In order to understand the orbital process involved in the ring-closure, let us consider the orbitals that change during the reaction. These are the four π-MOs of butadiene, which are transformed into two π- and two σ-orbitals in cyclobutene, as shown in Fig. 4.12.

This figure gives schematic diagrams of the MOs in which all contributions are treated as being from pure *p*-orbitals. They can be compared with the AM1-calculated MOs for *s-cis*-butadiene and cyclobutene. We must now decide how best to convert the relevant MOs of butadiene into those of cyclobutene in a continuous process. This is not difficult as we know that the two terminal CH$_2$-groups of the butadiene must rotate to make cyclobutene, but that the *two CH-groups remain planar throughout the reaction*. Therefore we must retain the orbital contributions for these two carbon atoms with the same relative phases as they have in butadiene in cyclobutene (because they cannot change their relative phases by rotating). This analysis implies that we must make doubly occupied orbitals from doubly occupied and unoccupied from unoccupied, as suggested by the orbital correlation diagram shown in Fig. 4.7. Therefore, the π-HOMO of cyclobutene (Ψ_{2C} in Fig. 4.12) must be derived from the lowest π-MO

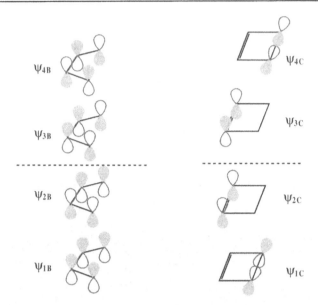

Fig. 4.12 The MOs of butadiene and cyclobutene that are interconverted during the ring-closure. The dashed lines indicate the border between occupied and unoccupied MOs

of butadiene (Ψ_{1B} in Fig. 4.12). Similarly, the π-LUMO of cyclobutene (Ψ_{3C}) must be derived from the highest π-MO of butadiene (Ψ_{4B}). Our analysis of the geometrical process required to complete the ring-closure is therefore reduced to converting Ψ_{2B} into Ψ_{1C} and Ψ_{3B} into Ψ_{4C}. If our analysis is correct, we will be able to use the same movement (either *conrotatory* or *disrotatory*) to perform both transformations.

In order to convert Ψ_{2B} into Ψ_{1C}, we can rotate the two CH2-groups in the same direction (*i.e.* perform a *conrotatory* ring-closure), as shown in Fig. 4.13.

Exactly the same movement also converts Ψ_{3B} into Ψ_{4C}, as shown in Fig. 4.14.

These *orbital correlations* give the following diagram for the thermal, *conrotatory* ring-closure reaction (Fig. 4.15).

This diagram corresponds to an allowed reaction, as shown in Fig. 4.7, but what would happen if we try to perform the *disrotatory* process? In this case, we must convert Ψ_{1B} into Ψ_{1C} and Ψ_{4B} into Ψ_{4C} (check these processes yourself), so that the π-MOs now correlate differently, as shown in Fig. 4.16 (see page 56).

Fig. 4.13 The *conrotatory* process that converts Ψ_{2B} into Ψ_{1C}

Fig. 4.14 The *conrotatory* process that converts Ψ_{3B} into Ψ_{4C}

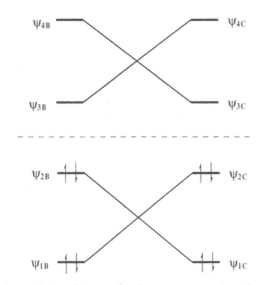

Fig. 4.15 Orbital correlation diagram for the *conrotatory* ring-closure

This diagram is representative of a forbidden reaction, as shown in Fig. 4.7. We can thus conclude that the thermal ring-closure of butadiene to cyclobutene is an allowed *conrotatory* process, but a forbidden *disrotato-*

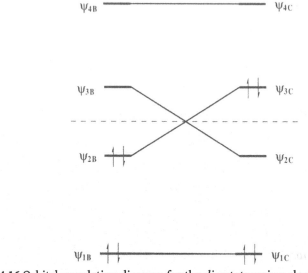

Fig. 4.16 Orbital correlation diagram for the *disrotatory* ring-closure

ry one. It should thus occur in a *conrotatory* fashion, as is observed experimentally.

The orbitals processes described above are quite independent of the electronic state of the reaction system, so can be used without change for the photochemical process. In this case, one electron is promoted from the HOMO to the LUMO of both butadiene and cyclobutene, so that, although the orbitals correlations are not changed, we now have three different types of MOs; doubly occupied, singly occupied and unoccupied. This means that, instead of the one border between doubly occupied and unoccupied MOs found for the thermal reaction, we now have two that may not be crossed by an orbital correlation line. This leads to the two diagrams shown in Fig. 4.17.

Now the *disrotatory* process is allowed and the *conrotatory* one forbidden – exactly the reverse of the thermal process. This result is quite general; photochemical electrocyclic reactions always occur in the opposite sense to their thermal equivalents.

We can generalise the results obtained above. Remember the rules outlined for determining the π-MOs of linear systems in Sect. 3.3.1. Remember also that *N* π-MOs are converted into *N-1* in an electrocyclic ring-closure. This means that, of all the π-MOs, only the "extra" one in the acyclic system, the HOMO, cannot be retained in the smaller, ring-closed π-sys-

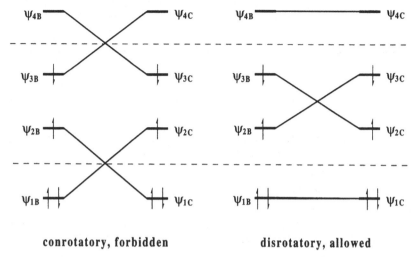

conrotatory, forbidden   **disrotatory, allowed**

Fig. 4.17 Orbital correlation diagrams for the photochemical ring-closure reaction: *conrotatory* (forbidden, left) and *disrotatory* (allowed, right)

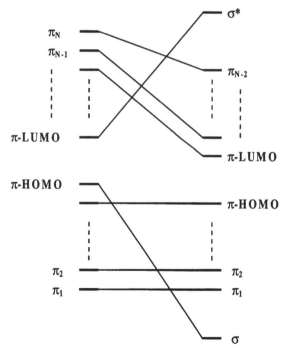

Fig. 4.18 Generic orbital correlation diagram for electrocyclic ring-closure reactions

Table 4.1 The Woodward-Hoffmann rules for electrocyclic reactions

Number of electrons	Thermal reaction	Photochemical reaction
4N	conrotatory	disrotatory
4N+2	disrotatory	conrotatory

tem. This is therefore the MO that we should use to form the new σ-MO. Thus a generic orbital correlation diagram (Fig. 4.18, see page 57) for such a process is as follows.

Although the orbitals crossing shown above are not real, this diagram gives us a convenient way to treat electrocyclic reactions. We can determine the nature of the process (*conrotatory* or *disrotatory*) by assuming that the π-HOMO must be converted to the new σ-MO. This was the hypothesis that was originally published by Woodward and Hoffmann. It means that, because the symmetry of the HOMO alternates as one more double bond (or two electrons) is added, the *conrotatory* or *disrotatory* nature of electrocyclic reactions also alternates. This can then be formulated as a set of rules, as shown in Table 4.1.

The same principles can now be used to describe other types of concerted reactions.

4.3.2
Cycloadditions

Cycloadditions are reactions in which two π-systems are added to each other so that two new σ-bonds are formed between their termini to give a new ring (Fig. 4.19).

The best known example, which we will treat here, is the Diels-Alder reaction, the prototype of which is the addition of ethylene to butadiene (Fig. 4.20).

There are two ways in which a chain can be added to a planar π-system. If the two new bonds are formed from the same side of the plane of the π-system, the addition is said to be *suprafacial*; an addition from opposite sides is *antarafacial* (Fig. 4.21).

As cycloadditions occur between two π-systems, each can undergo either *suprafacial* (*s*) or *antarafacial* (*a*) addition to give four different possibilities, *s+s, a+a, s+a* and *a+s*. The Woodward-Hoffmann rules for cycloadditions distinguish between these four modes according to the numbers of electrons involved.

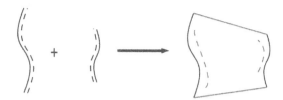

Fig. 4.19 Schematic representation of cycloaddtion reaction

Fig. 4.20 Diels-Alder cycloaddition between ethylene and butadiene

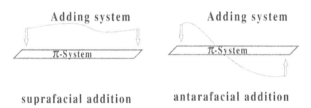

Fig. 4.21 *Suprafacial* (left) and *antarafacial* (right) addition to π-systems

For similar reasons to those outlined above for electrocyclic reactions, we only need to consider the frontier orbitals of the two systems. The HOMOs and LUMOs of butadiene and ethylene form the two new σ- and two new σ^*-MOs. These σ-MOs are grouped in bonding and antibonding combinations, as for cyclobutene above. The π-HOMO and π^*-LUMO of cyclohexene are derived directly from Ψ_{1B} and Ψ_{4B} of butadiene (Fig. 4.12). The two CH-groups of butadiene remain in the new π-system, as shown in Fig. 4.22.

The frontier orbitals of butadiene and ethylene and the σ-MOs of cyclohexene are shown schematically in Fig. 4.23. Once again, these schematic orbitals should be compared with those calculated with AM1 for *s-cis*-butadiene, ethylene and cyclohexene.

The two sets of orbitals can be divided into two classes according to whether they are symmetric or antisymmetric (in the sense that these terms

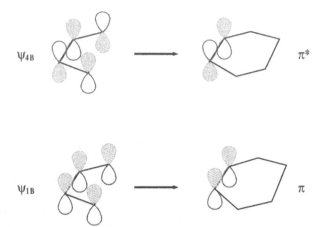

ψ_{4B} π^*

ψ_{1B} π

Fig. 4.22 The interconversion of the π-MOs in the Diels-Alder reaction

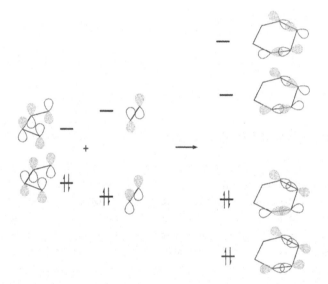

Fig. 4.23 The orbitals that form the new σ-bonds in the Diels-Alder reaction

are used in Sect. 3.3.1). This classification is indicated by the designations A and S in Fig. 4.23. There are two occupied orbitals on each side of the equation, one symmetric and one antisymmetric, and similarly, A and S unoccupied MOs. Furthermore, the HOMO of butadiene is antisymmetric, as is the LUMO of ethylene, whereas the LUMO of butadiene and the

HOMO of ethylene are symmetric. Thus, we can conclude that the two π-systems interact *via* two HOMO-LUMO two-electron attractive suprafacial interactions, as shown in Fig. 4.24. These two interactions need not be equally important. In most real Diels-Alder reactions, the olefin, or dienophile is substituted with electron-accepting groups. This sinks the energy level of its LUMO, making it a good electron acceptor. The interaction between the low-lying olefin LUMO and the diene HOMO then becomes the dominant stabilizing factor early in the reaction. However, it is possible to reverse this trend completely by using a very electron-rich olefin and a very electron-poor diene to give a so-called inverse Diels-Alder reaction. The orbitals obtained from the HOMO-LUMO interactions can be compared with the relevant σ-MOs of cyclohexene.

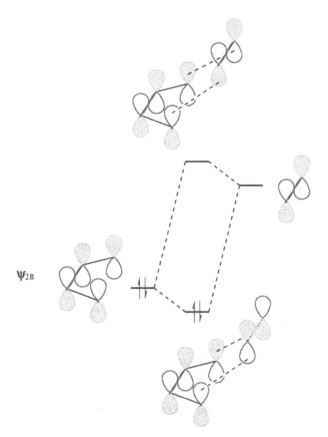

ψ_{2B}

Fig. 4.24a The interaction between the butadiene HOMO and the ethylene LUMO

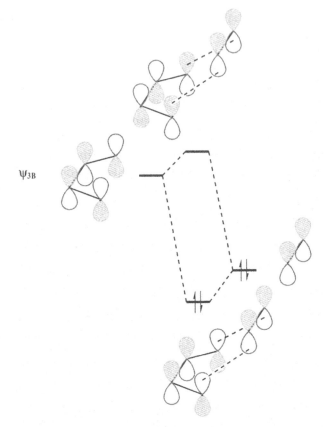

ψ_{3B}

Fig. 4.24b The interaction between the butadiene LUMO and the ethylene HOMO

Table 4.2 The Woodward-Hoffmann rules for cycloaddition reactions

Total number of π-electrons, I + J	Allowed photochemical reactions	Allowed thermal reactions
4N	*suprafacial(I), antarafacial(J)* *antarafacial(I), suprafacial(J)*	*suprafacial(I), suprafacial(J)* *antarafacial(I), antarafacial(J)*
4N + 2	*suprafacial(I), suprafacial(J)* *antarafacial(I), antarafacial(J)*	*suprafacial(I), antarafacial(J)* *antarafacial(I), suprafacial(J)*

Thus, the Diels-Alder reaction is a thermally allowed reaction in which the two components both add *suprafacially* and in which one has four π-electrons and one has two (a [4s + 2s] reaction). The Woodward-Hoffmann rules for cycloadditions with I π-electrons in one component and J in the other are summarised in Table 4.2.

Now, look at the molecular orbitals for the prototype Diels-Alder transition state in order to identify the six reacting orbitals outlined above.

4.3.3
Sigmatropic Rearrangements

Sigmatropic rearrangements are reactions in which a substituent that is hyperconjugating with a π-system migrates from one end to the other. In a sense, the bridged ethyl cation considered in Sect. 3.4.1. could be considered as the mid-point of a sigmatropic 1,2-hydrogen shift, but we will consider the model 1,3-hydrogen shift in propene. As for cycloadditions, sigmatropic rearrangements can either occur in a *suprafacial* manner, in which the shifting group remains on the same side of the π-system, or *antarafacially*, where the shifting group changes faces of the π-system (Fig. 4.25).

Sigmatropic reactions are best understood by considering their transition states, which are approximated as consisting of the two radicals obtained by breaking the bond to the migrating group. Thus, for the propene 1,3-hydrogen shift, a hydrogen atom must migrate across an allyl radical. The singly occupied MO (SOMO) of the allyl radical is shown schematically in Fig. 4.26 below.

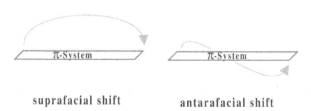

suprafacial shift   antarafacial shift

Fig. 4.25 *Suprafacial* (left) and *antarafacial* (right) addition to π-systems

Fig. 4.26 The SOMO of the allyl radical

The hydrogen atom, which has a singly occupied *s*-orbital, cannot shift *suprafacially* because at the transition state it would lie in the nodal plane of the allyl-SOMO and thus lose all bonding interaction (Fig. 4.27).

For the *antarafacial* shift, on the other hand, some overlap can be retained, even if the geometry of the transition state is very strained (Fig. 4.28).

Thus, the *suprafacial* shift is forbidden and the *antarafacial* one allowed. In fact, neither takes place without dissociation in propene itself because of the strain in the *antarafacial* transition state, but the fact that no *suprafacial* shift occurs is rationalised by the above arguments.

The Cope-rearrangement consists of a 1,3-shift of an allyl radical across the face of another (Fig. 4.29)

The two radical SOMOs can interact strongly in the *suprafacial* + *suprafacial* transition state, as shown in Fig. 4.30. (there are now two π-systems and so the same four stereochemical possibilities exist as for cycloadditions).

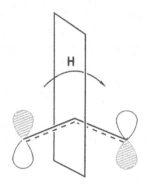

Fig. 4.27 Schematic transition state for the *suprafacial* 1,3-hydrogen shift in propene

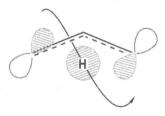

Fig. 4.28 Schematic transition state for the *antarafacial* 1,3-hydrogen shift in propene

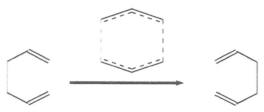

Fig. 4.29 Schematic representation of the Cope-rearrangement

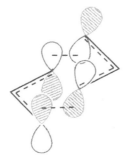

Fig. 4.30 The interaction between two allyl SOMOs in the Cope-rearrangement transition state

Table 4.3 The Woodward-Hoffmann rules for sigmatropic rearrangements

Total number of π-electrons, I + J	Allowed thermal shifts	Allowed photochemical shifts
4N	*suprafacial(I), antarafacial(J)* *antarafacial(I), suprafacial(J)*	*suprafacial(I), suprafacial(J)* *antarafacial(I), antarafacial(J)*
4N + 2	*suprafacial(I), suprafacial(J)* *antarafacial(I), antarafacial(J)*	*suprafacial(I), antarafacial(J)* *antarafacial(I), suprafacial(J)*

This reaction is an example of a 1,3-shift that is *suprafacial* for both components and involves two π-systems, each with 3 electrons. The MOs of 1,5-hexadiene and the Cope-rearrangement transition state show the reacting orbitals. Table 4.3 gives the Woodward-Hoffmann rules for sigmatropic rearrangements between π-systems with I and J electrons.

5 Elementary Symmetry

The aim of this chapter is to introduce symmetry elements, point groups and irreducible representations so that the symmetries of the individual orbitals can be understood. It is not intended to be a complete treatment of symmetry.

5.1
Symmetry elements

Symmetry elements are operations, such as rotation, reflection or inversion, that, when they are performed for a molecule or other object, give an object that is indistinguishable from the one on which the symmetry operation was performed. A perfect billiard ball (without any markings) can, for instance, spin without an observer being able to detect the movement. This is because a sphere has an infinite number of axes about which it can spin without changing its appearance. There are five common symmetry elements used to determine molecular symmetry.

5.1.1
The Identity, I

The identity is the simplest symmetry operation of all – we simply do nothing. Thus, the identity is the one symmetry element that is present in every object. It seems trivial even to consider a symmetry element that does absolutely nothing, but the identity is necessary for the accounting in symmetry treatments using *group theory*. It is given the symbol I (or sometimes E).

5.1.2
Proper Rotation Axes, C_n

Consider a perfect cylinder. Similarly to the billiard ball, if it were spinning about its axis, an observer would not be able to detect the movement. Put another way, we can rotate the cylinder by any angle about its axis without seeing any change. This is shown in the left-hand figure of Fig. 5.1. The cylinder can be rotated to any one of an infinite number of positions about its axis without any visible change. This means that the cylindrical axis is a rotation axis with an infinite number of possible rotation angles that give the same result. Rotation axes are denoted by the symbol c with a suffix to denote their *order*, or the number of equivalent positions obtained by rotation about the axis. Thus, the axis of the cylinder is denoted C_∞.

If we now draw a cross on the surface of the cylinder, as also shown in Fig. 5.1, we must rotate the cylinder by a full turn before it looks identical again. Of course, rotation by 360° is exactly equivalent to the identity, so that C_1 axes are not considered in molecular symmetry treatments. The third cylinder shown in Fig 5.1, however, now has two identical crosses drawn exactly at opposite sides of the surface of the cylinder. If we now rotate the cylinder by 180°, the crosses swap positions, but because they are identical the cylinder is indisinguishable from what we started with. There are thus two positions that are indistinguishable when the cylinder is rotated about its axis. The axis is therefore a two-fold proper rotation

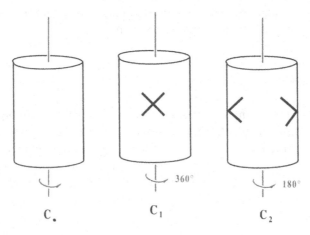

Fig. 5.1 Three different cylinders with inifinite-fold, one-fold and two-fold proper rotation axes

axis, or C_2. Quite generally, for a C_n axis, indistinguishable structures are obtained by rotating $(360/n)°$ about the axis.

Let us now apply the same principles to a molecule. Water has a C_2 axis, as shown in Fig. 5.2.

In the figure, the hydrogens have been labelled H_1 and H_2, but they are of course identical in a real molecule and therefore indistinguishable. Thus, rotating the black molecule by 180° about the C_2 axis (the vertical line) gives the right-hand water, which in real life, where the hydrogens are identical and all water molecules are the same, is identical with the starting structure.

Figure 5.3 shows the three-fold axis of ammonia.

This axis is exactly analogous to the two-fold axis of water except that we have three equivalent positions when we rotate the central ammonia in the direction shown by the lower arrow to give first the right-hand structure and, on a further rotation, a second structure in which H_2 takes the original position of H_1, H_1 that of H_3 and H_3 that of H_2 (Fig 5.4)

In Fig. 5.3, we can also rotate in the other (top arrow) direction about the C_3 axis. A single 120° rotation in this direction will give us the left-hand structure, which is indistinguishable from both the central and the right

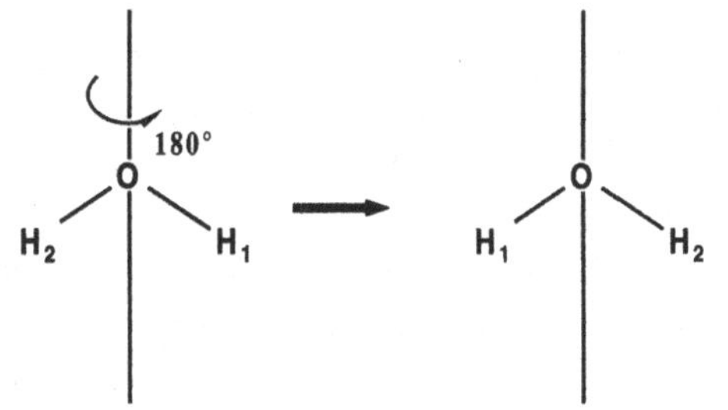

Fig. 5.2 The two-fold axis of water

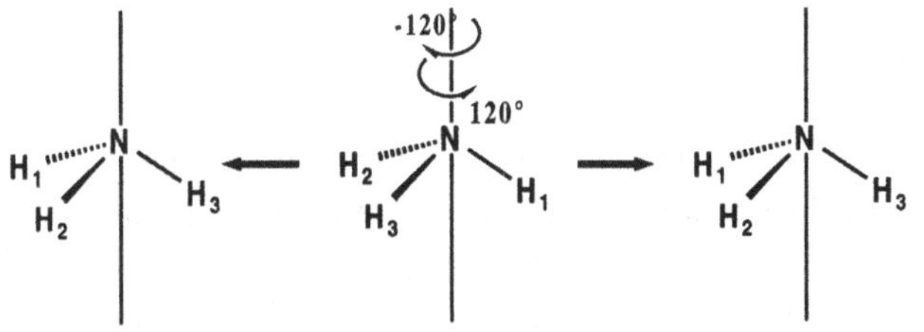

Fig. 5.3 Clockwise and anticlockwise rotations about the three-fold axis of ammonia

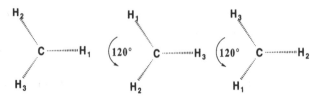

Fig. 5.4 Three-fold rotation axis of ammonia (projected on the plane of the page)

ammonia molecules. There is, however, an important difference between the C_2 axis of water and the C_3 of ammonia. If the hydrogens really were labelled, it would still be impossible to determine whether we had rotated water clockwise or anticlockwise to give the second equivalent structure. The result of the two rotations does not differ at all, and so they must be considered to be one symmetry operation (it doesn't matter how we got there, just what the result is). However, rotation in the two different directions in ammonia gives the right and left structures, which are distinguishable if we label the hydrogens. Therefore, the C_3 axis of ammonia counts not as one symmetry element, but two. 120° rotations in different directions are not exactly equivalent, so we really have one C_3 axis for each rotation direction. We will see later that ammonia has two C_3 axes that differ only in the direction of rotation. Such axes are called degenerate and point groups that contain them are also degenerate. If a molecule belongs to a degenerate point group, it can have degenerate orbitals or vibrations, as we saw above for cyclopropane, for instance. All C_n axes with n equal to or larger than three are degenerate.

The principal axis of any molecule or object is the proper rotation axis with the highest order n. If there are several proper rotation axes with the same order, the one that runs through the most atoms is the principal axis. Knowing which axis is the principal axis is important for determining the exact designation of mirror planes.

5.1.3
Mirror Planes, σ

The third type of symmetry element that we will consider is the mirror plane, usually designated σ. A mirror plane can be considered to be equivalent to an infinitely thin, double-sided planar mirror within the molecule or object. The simplest example of a mirror plane is the molecular plane for any planar molecule. Reflection of the atoms in the molecular plane

will leave them all unchanged. For instance, the plane of the page (the
molecular plane) is a mirror plane for the orientation of water shown in
Fig. 5.2. Because atoms that lie in a mirror plane are not moved by reflection in that plane, the molecular plane is a mirror plane for any planar molecule. In water, however, there is a second mirror plane perpendicular to
the molecular plane (Fig. 5.5).

Both mirror planes in water contain the C_2-principal axis and are therefore denoted vertical, or σ_v mirror planes. There are exceptions to this rule
for point groups with C_2-axes perpendicular to the principal axis. In this
case, vertical mirror planes that contain one of these perpendicular C_2-
axes are denoted σ_v and those that bisect the angle between two such axes
are denoted diagonal, or σ_d mirror planes. Planar BH_3, for instance, has
three σ_v planes (Fig. 5.6).

Perpendicular B_2H_4, on the other hand, has two σ_d mirror planes, as
shown in Fig. 5.7.

Note that the principal axis is determined by the fact that it runs
through the two boron atoms. The perpendicular C_2-axes bisect the angle

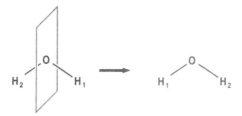

Fig. 5.5 Second mirror plane in water perpendicular to plane of page

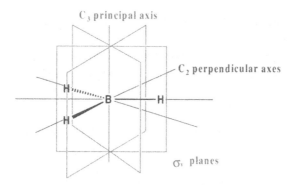

Fig. 5.6 σ_v mirrors planes in borane

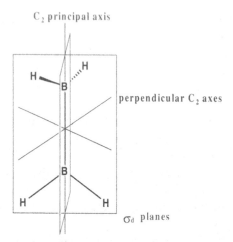

Fig. 5.7 σ_d mirror planes in B_2H_4

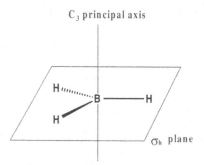

Fig. 5.8 σ_h mirror plane in BH_3

between the planes of the two BH_2 groups, whereas the two σ_d planes are identical with these planes.

The third type of mirror plane, the horizontal, lies perpendicular to the principal axis, as shown below in Fig. 5.8 for planar BH_3.

Mirror planes in objects without proper rotation axes are simply denoted σ.

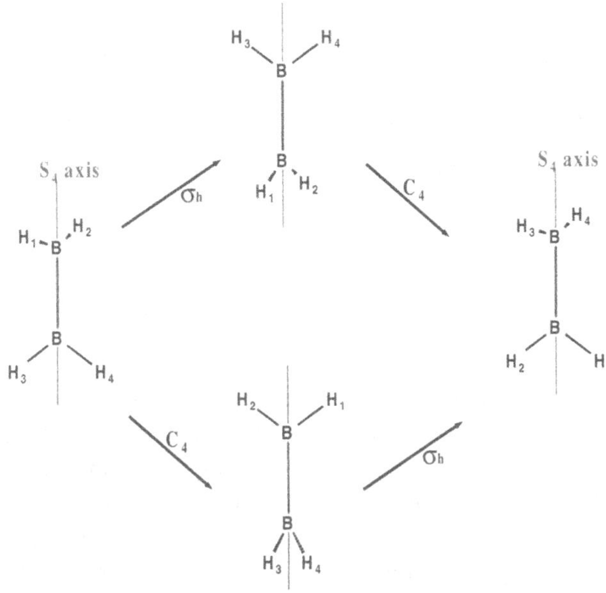

Fig. 5.9 Improper rotation axis in perpendicular B_2H_4

5.1.4
Improper Rotation Axes, S_n

Improper rotation axes are simply the combination of a proper rotation axis with a σ_h mirror plane. It does not matter in which order these two symmetry operations are applied, as shown in Fig. 5.9 for the S_4 axis of perpendicular B_2H_4.

Note that neither the C_4-axis nor the σ_h mirror plane are correct symmetry elements for the molecule. Only their combination is valid. Improper rotation axes with the order *2n* are often found to coincide with a proper rotation axis of order *n*.

5.1.5
Inversion Centres, i

Inversion centres are points in the centre of the molecule that act as infinitely small mirrors in every direction. This means that every atom is "reflected" to a position on the opposite side of the inversion centre (Fig. 5.10).

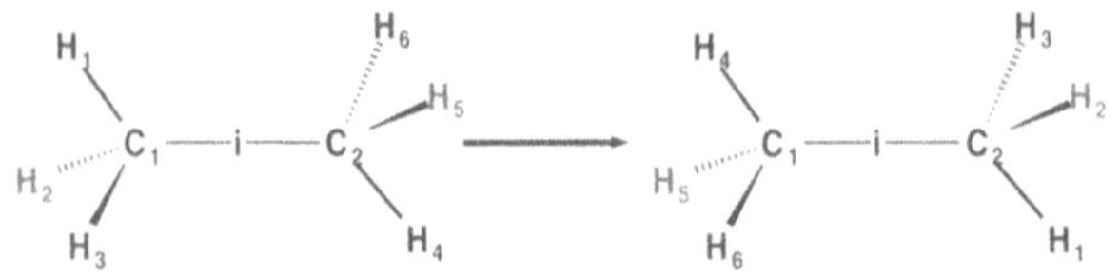

Fig. 5.10 Inversion centre in ethane bearing three different (labelled) substituents on each carbon atom

5.2
Point Groups

Point groups are mathematical entities defined by group theory that describe the complete symmetry of an object or moelcule (i.e. all its symmetry elements). It is necessary to determine the point group of a molecule in order to be able to use its symmetry to determine, for instance, the symmetries of the molecular vibrations or molecular orbitals. It is not, however, necessary to recognise all the symmetry elements of a molecule in order to be able to determine its point group. Figure 5.11 gives a simple scheme that allows the point group to be determined by recognising only key symmetry elements.

The most common point groups are C_1 (no symmetry at all) and C_s (just one mirror plane). The C_n, C_{nv} and C_{nh} point groups have just one proper rotation axis of order n and are distinguished from each other by their mirror planes. The D_n, D_{nh} and D_{nd} point groups are those with additional C_2-axes perpendicaular to the principal axis. The point groups of the molecule shown in the molecular orbital plots are given in the list of molecules in 6, so that you can try to use Fig. 5.11 to reproduce these point group assignments in order to familiarise yourself with the concept.

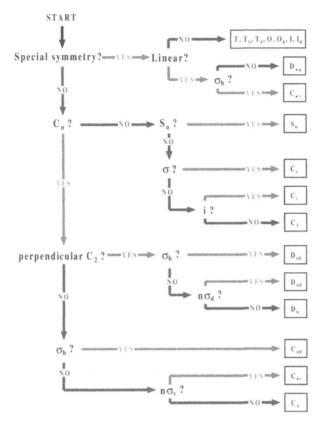

Fig. 5.11 A simple scheme for determining the point group of a molecule

5.3
Irreducible Representations and Character Tables

Irreducible representations, or *symmetry species,* are ways to depict all possible properties (such as molecular orbitals or normal vibrations) of a molecule in terms of their symmetry properties. The symmetry properties are defined in terms of the behaviour of the property when the symmetry elements of the molecular point group are applied. Thus, for instance, the σ_{CH}-MO of methylene (Fig. 5.12) is symmetrical with respect to rotation about the principal axis (it does not change) whereas the π_{CH2}-MO is antisymmetrical (it changes its phase, Fig. 5.13).

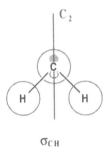

σ_{CH}

Fig. 5.12 Symmetrical behaviour of the methylene σ_{CH}-MO with respect to rotation about the principal axis

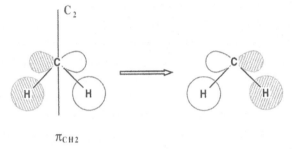

π_{CH2}

Fig. 5.13 Antisymmetrical behaviour of the methylene π_{CH2}-MO with respect to rotation about the principal axis

All possible properties can be described in terms of their behaviour (symmetrical or antisymmetrical) with respect to the symmetry elements of the molecule. The combinations of these behavioural patterns are the irreducible representations. They are classified using a symbolic notation that defines at least some of their behaviours. The meanings of the individual characters in the names of the irreducible representations are:

Letters:
   A Symmetrical with respect to a rotation about the principal axis
   B Antisymmetrical with respect to a rotation about the principal axis
   E Doubly degenerate
   T Triply degenerate

Subscripts:

1 Usually means symmetrical with respect to reflection in a σ_v plane, but the rules are more complicated if there are several different σ_v planes

2 Usually means antisymmetrical with respect to reflection in a σ_v plane, but the rules are more complicated if there are several different σ_v planes

g (gerade) symmetrical with respect to an inversion centre

u (ungerade) antisymmetrical with respect to an inversion centre

Superscripts:

$'$ symmetrical with respect to the single mirror plane in C_s, or to a σ_h plane

$''$ antisymmetrical with respect to the single mirror plane in C_s, or to a σ_h plane

These rules are often more complicated, especially for degenerate point groups, which will not be discussed here, but give a general idea of the principles.

The characteristics of a point group and its irreducible representations are usually collected in a *character table,* such as the one shown in Table 5.1 for C_{2v}.

The name of the point group is given in the top left corner. The symmetry elements are the column headers and the irreducible representations are given in the first column below the name of the point group. The characters are given in the body of the table and define whether the irreducible representation is symmetrical (+1) or antisymmetrical (−1) with respect to the symmetry element belonging to the column. Characters other than +1 and −1 can occur for degenerate point groups. The symbols to the right of the table give the irreducible representations of the molecular translation vectors (or dipole components) in the three Cartesian directions (x, y and z), those of the rotations about the Cartesian axes (R_x, R_y and R_z)

Table 5.1 Character table for the point group C_{2v}

C_{2v}	I	$C_2(z)$	$\sigma_v(xz)$	$\sigma_v(yz)$		
A_1	+1	+1	+1	+1	z	$\alpha_{xx}, \alpha_{yy}, \alpha_{zz}$
A_2	+1	+1	−1	−1	R_z	α_{xy}
B_1	+1	−1	+1	−1	x, R_y	α_{xz}
B_2	+1	−1	−1	+1	y, R_x	α_{yz}

and of the six independent components of the polarisabilty tensor (α_{xx}, α_{yy}, α_{zz}, α_{xy}, α_{xz}, and α_{yz}). These are important because, for instance, when we determine the normal vibrations of a molecule, we need to eliminate the rotations and translations. More importantly, however, vibrations with an irreducible representation corresponding to a translation vector are infrared active and those corresponding to one of the polarisability tensor components are Raman active. Thus, predictions about vibrational spectra can often be made simply from a knowledge of the irreducible representations of the normal vibrations. This and many other applications of symmetry are described in standard symmetry textbooks.

5.4
Degenerate Orbitals

Before we move on to the molecular orbital plots, we should consider degenerate MOs briefly. Confusion often arises because we have been taught, for instance, to expect the degenerate HOMOs of benzene to have nodal planes through two opposite carbon atoms and perpendicular thereto (Fig. 5.14).

However, if we use the full D_{6h} symmetry of the molecule, there are an infinite number of combinations of these two orbitals that all have the same energy and are equally good ways to represent the MOs. The HOMOs shown in Fig. 5.15 below are just one possibility.

Note that these MOs do not even look as if the have the correct symmetry for the molecule. This is because they must be considered as a degenerate pair and not individually. The sums of the two different depic-

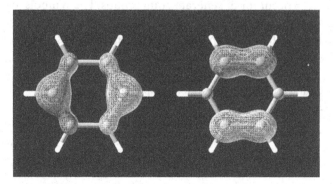

Fig. 5.14 HOMOs of benzene (optimised within D_{2h} symmetry)

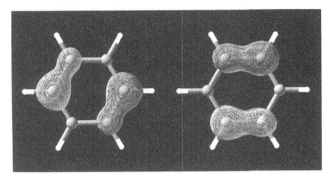

Fig. 5.15 HOMOs of benzene (optimised within D_{6h} symmetry)

tions of the degenerate MOs are identical. The MO plots in Fig. 5.14 have been made by slightly reducing the symmetry of the molecule to obtain the familiar orientations of degenerate MOs. In real calculations using the full symmetry, however, less recognisable forms are usually found.

6 Glossary

3D-electron density contour diagram

Diagrams of the electron density (the square of the wavefunction) in a single MO or complete molecule contoured at an appropriate electron density level to visualise the "shape" of the orbital or total electron density of the molecule.

Allowed, forbidden reactions

Allowed and forbidden reactions were defined by Woodward and Hoffman as those that proceed without and with a change in the orbital occupancies, respectively. This simple picture is only strictly applicable within *point groups* that do not allow the two orbitals to interact, but is still a good interpretative framework.

Anomeric effect

The stabilising interaction of a lone pair donor orbital on a heteroatom (usually oxygen) with a σ^*-orbital of an α, β-bond to an electronegative element (again, usually oxygen). The anomeric effect is equivalent to *negative hyperconjugation* and the reverse of *hyperconjugation*.

Atomic Orbital (AO)

Strictly speaking, atomic orbitals are only applicable to the atom itself. Within the *LCAO-approximation*, however, AOs are combined to make up the *molecular orbitals, MOs*.

"Banana" bond

"Banana", or bent, bonds are found in strained compounds, especially cyclopropanes, where they are caused by the *Walsh orbitals*. They are characterised by the fact that, in contrast to normal σ-bonds, the centre of the electron density associated with the orbital lies outside the line connecting the two atoms.

Character table

Table of *symmetry elements, irreducible representations*, characters of the individual irreducible representations and assignments of vector and tensor properties for a given point group.

Conrotatory, disrotatory	In an electrocyclic reaction, the conrotatory process is the one in which the two end groups that rotate to form the new bond do so in the same direction. For the disrotatory process, they rotate in opposite directions.
Degenerate orbitals	A set of two molecular orbitals belonging to an E-, three belonging to a T-, four belonging to a G- or five belonging to an H-*irreducible representation*. The orbitals in a degenerate set have exactly the same energy and can be combined to give an alternative representation of the degenerate set with no energetic consequences.
Dipole moment	The dipole moment, μ, of two equal charges, $\pm Q$, of opposite sign at a distance r is given by:

$$\mu = Q \cdot r$$

For an array of charges, such as a molecule consisting of i atoms each with a net charge Q_i

$$\mu = \sum_i Q_i \cdot r_i$$

This definition can be extended to the nuclear charges and electron density of the molecule within a quantum mechanical treatment.

Electrocyclic reaction	An electrocyclic reaction is one in which a π-system consisting of n π bonds interconverts with a cyclic system with $(n-1)$ π-bonds and one extra π-bond.
Electronegativity, electropositive	The electronegativity of an element is its tendency to accumulate electrons (negative charge). The higher the electronegativity, the more negative the element is likely to be in a molecule. There are many different electronegativity scales. Electropositive elements are those (such as metals) with low electronegativities.
Frontier orbital theory	The frontier orbitals are the *HOMO* and *LUMO*. Frontier orbital theory uses only these orbitals to treat reactivity problems by assuming that the two-electron interaction between HOMOs and LUMOs is likely to be the strongest between two interacting molecules. This is justified by the fact that the energy gap between HOMO and LUMO is smaller than between any other occupied and virtual orbitals.

Global minimum   Depending on the definition of a molecule – either as a collection of atoms with a given total formula or as a given bonding pattern – the global minimum is the most stable minimum energy structure that exists. Finding the gobal minimum for large molecules is a largely unsolved problem. Less stable minima than the global minimum are known as *local minima*.

Group theory   A group is defined as a collection of elements (in the case of symmetry groups, symmetry operations) that are related to each other by a given set of rules. Group theory allows us to apply manipulations such as multiplication, definition of subgroups and classes, etc.

Group orbital   A group orbital is an orbital that is assigned to a given fragment or group within a molecule. Usually group orbitals resemble the *MOs* of the given fragment closely. MOs for larger molecules can be built up qualitatively as linear combinations of group orbitals.

**Highest Occupied   The HOMO is simply the doubly occupied orbital with
Molecular Orbital   the highest energy. It is one of the *frontier orbitals*.
(HOMO)**

Hückel theory   Hückel theory was the first *LCAO*-MO approach to calculating the electronic structure of molecules. Hückel theroy considers only the p-orbitals perpendicular to a planar π-system, does not include electron-electron repulsion and assumes that all bonding overlaps are equal and all nonbonding zero.

Hyperconjugation   Hyperconjugation is the stabilising interaction between an occupied σ-MO and an empty acceptor orbital on the adjacent atomic center. It is the reverse of the *anomeric effect*.

**Irreducible represen-   Within the context of the symmetry groups discussed
tation, symmetry   here, irreducible representations are combinations of
species**   symmetry elements that are orthogonal to each other and describe the symmetry species (e. g. vibrations, orbitals, etc.) of the *point group*. As the name implies, the irreducible representations cannot be reduced to simpler ones.

Koopmans' theorem	Koopmans' theorem is that the ionisation potential of a molecule is approximately minus the energy of its *HOMO*.
Lewis structure	The Lewis structure of a molecule is its bonding pattern described in terms of the octet rule with formal charges, single, double and triple bonds.
Lewis acid	A Lewis acid is a compound with a low-lying *LUMO* that allows it to accept electrons from *Lewis bases*.
Lewis base	A Lewis base is a compound with a high-lying *HOMO* that allows it to donate electrons to *Lewis acids*.
Linear Combination of Atomic Orbitals (LCAO)	The LCAO approximation allows us to build up the *MOs* of a molecule as linear combinations of a basis set of *atomic orbitals* assigned to each atom. The approximation provides a simple interpretative framework for the nature of MOs and also has important computational advantages.
Lowest Unoccupied Molecular Orbital (LUMO)	The LUMO is simply the virtual (unoccupied) *MO* with the lowest energy. It is one of the *frontier orbitals*.
Mirror plane	Mirror planes (denoted σ) are imaginary planes through a molecule that, when treated as infinitely thin two-sided mirrors, reflect the molecule to give a new structure that is indistinguishable from the starting one.
Molecular Orbital (MO)	Within the context of this book, it is sufficient to define an MO as an orbital belonging to the entire molecule in question. Within the *LCAO approximation*, these MOs are expressed in terms of contributions from individual *AOs*.
Negative hyper-conjugation	Negative hyperconjugation is another name for the *anomeric effect*. It is usually used in connection with anions.
Net atomic charge	Net atomic charges have no physical significance. They result from some scheme for partitioning the electrons in the molecule to the different atoms and are typically reported in the *population analysis*.

Normalisation, normalised	Normalisation is the mathematical process by which the total probablility, for instance of finding an electron in an orbital, is adjusted to be unity.
Orbital correlation diagram	For concerted reactions, such as those treated by the Woodward-Hoffmann rules, an orbital correlation diagram traces the changes in a given *MO* from starting point to product.
Orbital interaction diagram	For bimolecular reactions, orbital interaction diagrams visualise the mixing of the orbitals of the two reacting molecules. Orbital interaction diagrams are often limited to the *frontier orbitals*.
Orthogonal	Two species are orthogonal when they cannot interact with each other (i.e. their *overlap* is zero).
Overlap	Within the current context, overlap is the interaction between orbitals. Bonding overlap stabilises, antibonding destabilises.
Point group	Point groups describe the symmetry characteristics of molecules in terms of their *symmetry elements*.
Population analysis	A population analysis is a (largely artificial) partitioning of the electrons in the molecule to the individual atoms in order to obtain *net atomic charges*, bond orders, etc.
Singly Occupied Molecular Orbital (SOMO)	For radicals, triplets, etc. The SOMO(s) are the orbitals that formally contain only one electron.
Strain, strain energy	A molecule is strained when the sum of the dissociation energies of its bonds is smaller than the corresponding sum of an arbitrary unstrained reference molecule. For instance, the bond dissociations energies of the three CC-bonds in cyclopropane are smaller than three times the "normal" CC-bond dissociation energy. Strain leads to increased reactivity and decreased stability.
Suprafacial, antarafacial	A suprafacial process is one in which two new bonds are made to the same face of a π-system, or in which a bond is broken and a new one made on the same face. Antarafacial processes involve the two different faces of the π-system.

Symmetry element A symmetry element is an operation, such as a *mirror plane*, that transforms a molecule into a structure that is indistinguishable from the starting one.

Symmetry-adapted combination If two atoms, for instance, are symmetrically equivalent, their individual *AOs* do not have the correct symmetry characteristics to belong to one of the *irreducible representations* of the *point group*. In this case, the AOs must be combined to give symmetry-adapted combinations.

Valence bond theory Valence bond (or VB) theory is an alternative approach to MO-theory in which resonance hybrids between different *Lewis structures* can be formed.

Virtual orbital A virtual orbital is simply an unoccupied one.

Walsh orbital The Walsh orbitals of cyclopropane are its degenerate *HOMOs*. They are largely responsible for the *strain* and for the *"banana" bond* character of the cyclopropane CC-bonds.

Walsh diagram Walsh diagrams are a simple MO-technique used to rationalise the structure of simple molecules and fragments by considering the energy changes in the *MOs* when the structure is distorted.

7 List of Molecules

Name	Formula	Point group	Electrons	HOMO	SOMO	Heat (kcal/mol)
Hydrogen	H_2	$D_{\infty h}$	2	1	0	-41.76
Methylene (triplet)	CH_2	C_{2v}	6	2	2	82.09
Amino cation (triplet)	NH_2^+	C_{2v}	6	2	2	279.60
Methylene (singlet)	CH_2	C_{2v}	6	3	0	110.85
Borane (3)	BH_3	D_{3h}	6	3	0	26.26
Methyl radical (planar)	CH_3	D_{3h}	7	3	1	31.25
Borane (3) radical anion	BH_3^-	C_{3v}	7	3	1	49.96
Ammonia radical cation	NH_3^+	D_{3h}	7	3	1	209.50
Methane	CH_4	T_d	8	4	0	-8.79
Ammonia	NH_3	C_{3v}	8	4	0	-7.30
Water	H_2O	C_{2v}	8	4	0	-59.25
Hydrogen fluoride	HF	$C_{\infty v}$	8	4	0	-74.28
Protonated methane	CH_5^+	C_1	8	4	0	224.35
Acetylene	$HCCH$	$D_{\infty h}$	10	5	0	54.78
Vinyl cation	$C_2H_3^+$	C_{2v}	10	5	0	261.43
Hydrogen cyanide	HCN	$C_{\infty v}$	10	5	0	30.99
Carbon monoxide	CO	$C_{\infty v}$	10	5	0	-5.70
Nitrogen	N_2	$D_{\infty h}$	10	5	0	11.14
Nitrogen monoxide	NO	$C_{\infty v}$	10	5	0	1.18
Ethylene	C_2H_4	D_{2h}	12	6	0	16.45
Methylene imine	CH_2NH	C_s	12	6	0	18.66
Formaldehyde	$HCHO$	C_{2v}	12	6	0	-31.51
Diimide	N_2H_2	C_{2h}	12	6	0	31.52
Diborane	B_2H_4	D_{2h}	12	6	0	5.29
Oxygen (triplet)	O_2	$D_{\infty h}$	12	6	0	0.70
Ethyl cation (bisected)	$C_2H_5^+$	C_s	12	6	0	216.76

Name	Formula	Point group	Electrons	HOMO	SOMO	Heat (kcal/mol)
Ethyl cation (eclipsed)	$C_2H_5^+$	C_s	12	6	0	216.78
Ethyl cation (bridged)	$C_2H_5^+$	C_{2v}	12	6	0	226.26
Ethyl radical (bisected)	C_2H_5	C_s	13	6	1	18.41
Ethyl radical (eclipsed)	C_2H_5	C_s	13	6	1	18.42
Ethane (staggered)	C_2H_6	D_{3d}	14	7	0	-17.44
Ethane (eclipsed)	C_2H_6	D_{3h}	14	7	0	-16.19
Methylamine	CH_3NH_2	C_s	14	7	0	-7.41
Methanol	CH_3OH	C_s	14	7	0	-57.05
Methyl fluoride	CH_3F	C_{3v}	14	7	0	-61.04
Hydrazine	N_2H_4	C_{2h}	14	7	0	13.65
Hydrogen peroxide	H_2O_2	C_2	14	7	0	-35.35
Fluorine	F_2	$D_{\infty h}$	14	7	0	-22.48
Cyclopropenium cation	$C_3H_3^+$	D_{3h}	14	7	0	276.36
Borane:ammonia complex	$BH_3{:}NH_3$	C_{3v}	14	7	0	-19.30
Acetonitrile	CH_3CN	C_{3v}	16	8	0	21.34
Methyl isocyanide	CH_3NC	C_{3v}	16	8	0	50.86
Allene	H_2CCCH_2	$S4$	16	8	0	49.39
Ketene	H_2CCO	C_{2v}	16	8	0	-5.69
Diazomethane	H_2CN_2	C_{2v}	16	8	0	62.59
Carbodiimide	CH_2N_2	D_{2h}	16	8	0	41.37
Carbon dioxide	CO_2	$D_{\infty h}$	16	8	0	-79.86
Cyclopropene	C_3H_4	C_{2v}	16	8	0	74.78
Diazirine	CH_2N_2	C_{2v}	16	8	0	86.80
Allyl cation	$C_3H_5^+$	C_{2v}	16	8	0	226.18
Propyne	C_3H_4	C_{3v}	16	8	0	43.38
Trimethylene (edge-on)	C_3H_6	C_{2v}	18	8	2	49.25

Name	Formula	Point group	Electrons	HOMO	SOMO	Heat (kcal/mol)
Propene	C_3H_6	C_s	18	9	0	6.54
Acetaldehyde	CH_3CHO	C_s	18	9	0	-41.60
Formamide	$HCONH_2$	C_s	18	9	0	-44.79
Formic acid	$HCOOH$	C_s	18	9	0	-97.41
Formyl fluoride	$HCOF$	C_s	18	9	0	-92.96
Nitrosomethane	CH_3NO	C_s	18	9	0	-1.94
Ozone	O_3	C_{2v}	18	9	0	37.70
Cyclopropane	C_3H_6	D_{3h}	18	9	0	17.75
Aziridine	C_2H_5N	C_s	18	9	0	33.09
Oxirane	C_2H_4O	C_{2v}	18	9	0	-8.99
n–Propyl cation	$C_3H_7^+$	C_s	18	9	0	211.69
Propane	C_3H_8	C_{2v}	20	10	0	24.30
Dimethyl ether	C_2H_6O	C_{2v}	20	10	0	-53.21
Ethyl fluoride	C_2H_5F	C_s	20	10	0	-66.33
Cyclobutadiene	C_4H_4	D_{2h}	20	10	0	111.22
Cyclopropenone	C_3H_2O	C_{2v}	20	10	0	49.55
Ethanol	C_2H_5OH	C_s	20	10	0	-62.70
Tetrahedrane	C_4H_4	T_d	20	10	0	159.20
Fluoroethyl anion	$C_2H_5F^-$	C_s	20	10	0	-23.58
trans–1,3–Butadiene	C_4H_6	C_{2h}	22	11	0	29.87
cis–1,3–Butadiene	C_4H_6	C_{2v}	22	11	0	30.65
trans–Acrolein	C_3H_4O	C_s	22	11	0	-16.56
cis–Acrolein	C_3H_4O	C_s	22	11	0	-16.34
trans–Glyoxal	$C_2H_2O_2$	C_{2h}	22	11	0	-58.75
cis–Glyoxal	$C_2H_2O_2$	C_{2v}	22	11	0	-56.28
Methylazide	CH_3N_3	C_s	22	11	0	76.61

Name	Formula	Point group	Electrons	HOMO	SOMO	Heat (kcal/mol)
Methylenecyclopropane	C_4H_6	C_{2v}	22	11	0	47.62
Cyclopropanone	C_3H_4O	C_{2v}	22	11	0	3.16
Cyclobutene	C_4H_6	C_{2v}	22	11	0	45.72
Bicyclobutane	C_4H_6	C_{2v}	22	11	0	78.02
Cyclopropylcarbinyl cation (bisected)	$C_4H_7^+$	C_s	22	11	0	232.39
Cyclopropylcarbinyl cation (perpenticular)	$C_4H_7^+$	C_s	22	11	0	245.76
S_N2 TS	$CH_3F_2^-$	D_{3h}	22	11	0	-38.81
Fluoride:fluoromethane complex	$CH_3F_2^-$	C_{3v}	22	11	0	-67.38
trans–2–Butene	C_4H_8	C_{2h}	24	12	0	-3.39
Acetone	C_2H_6O	C_{2v}	24	12	0	-49.24
Isopropenol	C_2H_6O	C_s	24	12	0	-39.56
Nitromethane	CH_3NO_2	C_s	24	12	0	-9.99
Cyclobutane	C_4H_8	D_{4h}	24	12	0	-1.03
Guanidinium cation	$CH_6N_3^+$	D_{5h}	24	12	0	151.10
Cyclopentadiene	C_5H_6	C_{2v}	26	13	0	37.01
Bicyclo[2,1,0]–2–pentene	C_5H_6	C_s	26	13	0	98.38
Pyrrole	C_4H_5N	C_{2v}	26	13	0	39.82
Furane	C_4H_4O	C_{2v}	26	13	0	2.90
Cyclopentadienyl anion	$C_5H_5^-$	D_{5h}	26	13	0	25.12
Bicyclo[1,1,1]–propellane	C_5H_6	D_{3h}	26	13	0	188.58
Pentadienyl radical	C_5H_7	C_{2v}	27	13	1	49.61
p–Benzyne	C_6H_4	C_{2v}	28	13	2	134.26
m–Benzyne	C_6H_4	C_{2v}	28	13	2	133.03
o– Benzyne	C_6H_4	C_{2v}	28	13	2	135.97
Cyclopentene	C_5H_8	C_{2v}	28	14	0	2.92
Bicyclo[1,1,1]–pentane	C_5H_8	D_{3h}	28	14	0	82.90

Name	Formula	Point group	Electrons	HOMO	SOMO	Heat (kcal/mol)
Spiropentane	C_5H_8	C_s	28	14	0	50.39
Cyclopentane	C_5H_{10}	C_2	30	15	0	-28.85
Benzene	C_6H_6	D_{6h}	30	15	0	21.96
Dewar benzene	C_6H_6	C_{2v}	30	15	0	116.67
Pyridine	C_5H_5N	C_{2v}	30	15	0	34.32
Pyrazine	$C_4H_4N_2$	C_{2v}	30	15	0	47.37
Cyclopentadienone	C_5H_4O	C_{2v}	30	15	0	20.60
2,2-Dimethyloxirane	C_4H_8O	C_s	30	15	0	-20.53
Benzene (optimised in D_{6h} symmetry)	C_6H_6	D_{6h}	30	15	0	21.96
1,3,5-Hexatriene	C_6H_8	C_{2h}	32	16	0	42.85
Bicyclo[2,1,1]-2-hexene	C_6H_8	C_{2v}	32	16	0	75.45
NF$_3$O	F_3NO	C_{3v}	32	16	0	-14.65
Cyclohexene	C_6H_{10}	C_2	34	17	0	-10.12
1,5-Hexadiene	C_6H_{10}	C_1	34	17	0	19.55
Cope rearrangement TS	C_6H_{10}	C_{2h}	34	17	0	58.30
Diels–Alder reaction TS	C_6H_{10}	C_s	34	17	0	70.09
Cyclohexane	C_6H_{12}	D_{3d}	36	18	0	-38.61
Norbornadiene	C_6H_{12}	C_{2v}	36	18	0	67.62
Maleic anhydrate	$C_4H_2O_3$	C_{2v}	36	18	0	-76.43
Aniline	$C_6H_5NH_2$	C_{2v}	36	18	0	21.41
Quadricyclane	C_7H_8	C_{2v}	36	18	0	104.27
Cycloheptatriene anion	$C_7H_7^-$	D_{7h}	36	18	0	57.04
Cyclooctatetraene (planar)	C_8H_8	D_{8h}	40	20	0	87.19
2-Hydroxypyrane (axial)	$C_5H_{10}O_2$	C_1	42	21	0	-119.31
2-Hydroxypyrane (equatorial)	$C_5H_{10}O_2$	C_1	42	21	0	-118.66
Nitrobenzene	$C_6H_5NO_2$	C_{2v}	46	23	0	25.18

Name	Formula	Point group	Electrons	HOMO	SOMO	Heat (kcal/mol)
Cyanide anion	CN^-	$C_{\infty v}$	10	5	0	44.01
Thiocyanate anion	SCN^-	$C_{\infty v}$	16	8	0	-14.20
Azide anion	N_3^-	$D_{\infty h}$	16	8	0	57.83
Nitro anion	NO_2^-	C_{2v}	18	9	0	-68.65
Carbonate dianion	CO_3^{2-}	D_{3h}	24	12	0	-48.88
Imidazolate	$C_4H_4N^-$	C_{2v}	26	13	0	28.16
Dimethylsulfoxide	C_2H_6OS	C_s	26	13	0	-39.39
Ethylendiamine (en)	$C_2H_8N_2$	C_2	26	13	0	-2.85
Glycine	$C_2H_5NO_2$	C_1	30	15	0	-98.05
1,3–Diaminopropane (tn)	$C_3H_{10}N_2$	C_1	32	16	0	-6.79
Oxalate dianion	$C_2O_4^{2-}$	D_2	34	17	0	-129.81
1,2–Dithiolene	$C_4H_6S_2$	C_s	34	17	0	50.85
Acetylacetate	$C_5H_7O_2^-$	C_{2v}	40	20	0	-105.55
Diethylentriamine (dien)	$C_4H_{13}N_3$	C_s	44	22	0	8.57
1,3,5–Triaminocyclohexane (tach)	$C_6H_{15}N_3$	C_{3v}	54	27	0	-0.05
1,4,7–Triazacyclononane (tacn)	$C_6H_{15}N_3$	C_3	54	27	0	27.03
3,3'–Iminobispropylamine (dpt)	$C_6H_{17}N_3$	C_1	56	28	0	-3.36
2,2'–Bipyridyl	$C_{10}H_8N_2$	C_{2v}	58	29	0	75.40
Triethylentetramine (trien)	$C_6H_{18}N_4$	C_1	62	31	0	18.33
1,4,7,10–Tetraazacyclododecane (cyclen)	$C_8H_{20}N_4$	C_2	72	36	0	41.98
Tris(2–aminoethyl)amine (tren)	$C_8H_{22}N_4$	C_1	74	37	0	12.90
Triaminomethylethane (tame)	$C_{10}H_{25}N_3$	C_1	80	40	0	-10.59
1,4,8,11–Tetraazacyclo-tetradecane (cyclam)	$C_{10}H_{24}N_4$	C_1	84	42	0	26.56
1,4,8,11–Tetrathiacyclotetradecane	$C_{10}H_{20}S_4$	C_1	84	42	0	-0.54
2,2':6',2''–Terpyridyl	$C_{15}H_{11}N_3$	C_{2v}	86	43	0	120.10
Ethylenediamintetraacetate (edta)	$C_6H_4N_2O_8^{4-}$	C_{2v}	90	45	0	-34.42

Name	Formula	Point group	Electrons	HOMO	SOMO	Heat (kcal/mol)
18-Crown-6	$C_{12}H_{24}O_6$	C_2	108	54	0	-234.79
6,13-Diamino-6,13-dimethyl-1,4,8,11-tetraazacyclotetradecane (diammac)	$C_{12}H_{30}N_6$	C_1	108	54	0	38.74
1,4,7,10,13,16-Hexathiacyclooctadecane	$C_{12}H_{24}S_6$	C_2	108	54	0	10.77
1,3,6,8,10,13,16,19-Octa-azabicyclo [6.6.6]icosane (sep)	$C_{12}H_{30}N_8$	C_2	118	59	0	117.02
1,4,8,11,15,18-Hexa-azabicyclo [6.6.6]icosane (sar)	$C_{14}H_{32}N_6$	C_1	118	59	0	66.09
1,4,7,10-Tetraacetato-1,4,7,10-tetraazacyclododecane (dota)	$C_{12}H_{16}N_4O_8^{\,4-}$	C_2	136	68	0	-39.38
1,2-Bis(diphenylphosphino)ethane	$C_{26}H_{24}P_2$	C_{2v}	138	69	0	135.10

8 General Information

The Chemist's Electronic Book of Orbitals introduces students and researchers to the world of molecular orbitals using three-dimensional VRML representations. The electronic components of the book are delivered on the accompanying CD-ROM. The printed part gives a short introduction into basic qualitative molecular orbital theory to enable the reader to make the most of the VRML orbitals on the CD. The CD itself contains an extended interactive textbook and a broad selection of organic compounds and ligands for inorganic complexes including their molecular orbitals. In several demonstrations, the student can change parameters and observe the change in the orbitals or zoom and rotate the 3D-VRML objects.

The CD-ROM is intended for use on a large variety of platforms including Windows 95/98/NT-based PCs, Macs, and UNIX-based systems. It only requires a CD-ROM drive, a 4th generation Web browser with Java and JavaScript capabilities (Netscape 4 or MicroSoft Internet Explorer 4), a VRML plug-in, a high resolution graphics card (we recommend a color depth of 16 bits, corresponding to 16.7 million colors), and a pointing device such as a mouse, trackball or touchpad.

Installation

The book is accessible directly from CD-ROM. However, for improved performance, it is useful to copy the contents to a local hard disk. This will need about 450 MB of free disk space.

For the installation of the Web browser, please refer to the installation instructions. Some browsers already contain a VRML plug-in, for example the Netscape Communicator Professional Version 4.0x includes Silicon Graphics´ CosmoPlayer. We have tested the CD-ROM with both Netscape browsers and MS Internet Explorer and several plug-ins. Good results are given by the CosmoPlayer for Netscape and Microsoft's own VRML viewer for Internet Explorer. They are available via the Internet at:

CosmoPlayer (Windows95/98/NT, Mac OS 7.6.1 or higher,
and IRIX 5.3 or higher):
http://cosmosoftware.com/

MS VRML 2.0 viewer (only Windows95/98/NT):
http://www.microsoft.com/vrml/toolbar/
http://www.eu.microsoft.com/vrml/toolbar/

Another Web site is Intervista Software at http://www.intervista.com/, they also
offer a free Mac (Netscape) and PC (Netscape and MS IE) viewer. A more general
web site about VRML which contains many links and information on VRML view-
ers and plug-ins is "The VRML Repository" (http://www.sdsc.edu/vrml/reposito-
ry.html). Here you can also find a list of plug-ins for other platforms.

These URLs may have already changed by the time you buy this book, so we
cannot guarantee that these plug-ins can still be found at the above locations.

For the installation of the VRML plug-ins, again please refer to the installation
instructions of the plug-in.

Use

In order to start *The Chemist's Electronic Book of Orbitals* on a PC running Windows 95/98/NT, open the Windows *Explorer*, change to the drive letter of your CD-
ROM and double-click on the file *index.htm* in the root directory of your CD drive.
If your Web browser is installed correctly, it will open and display the main page
of the book. Alternatively, start your browser and open *index.htm* from the root
directory of your CD drive. The latter procedure will work on other platforms as
well (Mac and UNIX).

Every page of the book has a navigation menu in the left, yellow column. The
current page is highlighted with a black square. Move your mouse and the next
choice is also highlighted with a black square. Click on *Contents* and a list of all
linked pages appears in the main window. From here, you can browse through the
text, go to a specific chapter or view an interactive demonstration. There is also a
Glossary which explains many terms used within the textbook. These expressions
are also linked from within the electronic version. If you want to go directly to the
VRML orbitals, click on *Molecules in VRML*. You will see an alphabetic list of all
molecules on the CD. A more detailed list is also available from the page. In order
to select this list, choose *General Information – Installation – Use*.

What can you do with all these VRML files on the CD-ROM? They are stored as
3D-objects, which can be rotated, observed from all sides and zoomed at will.

If you decide to use SGI's CosmoPlayer, it offers a very nice help (click on the
"?" in the lower right corner). This explains every button of the VRML plug-in,
gives a quick reference and helps you set user-defined preferences.